Studies in Systems, Decision and Control

Volume 229

The series "Studies in Systems, Decision and Control" (SSDC) covers both new developments and advances, as well as the state of the art, in the various areas of broadly perceived systems, decision making and control–quickly, up to date and with a high quality. The intent is to cover the theory, applications, and perspectives on the state of the art and future developments relevant to systems, decision making, control, complex processes and related areas, as embedded in the fields of engineering, computer science, physics, economics, social and life sciences, as well as the paradigms and methodologies behind them. The series contains monographs, textbooks, lecture notes and edited volumes in systems, decision making and control spanning the areas of Cyber-Physical Systems, Autonomous Systems, Sensor Networks, Control Systems, Energy Systems, Automotive Systems, Biological Systems, Vehicular Networking and Connected Vehicles, Aerospace Systems, Automation, Manufacturing, Smart Grids, Nonlinear Systems, Power Systems, Robotics, Social Systems, Economic Systems and other. Of particular value to both the contributors and the readership are the short publication timeframe and the world-wide distribution and exposure which enable both a wide and rapid dissemination of research output.

Indexed by SCOPUS, DBLP, WTI Frankfurt eG, zbMATH, SCImago.

All books published in the series are submitted for consideration in Web of Science.

Frank Ehlers

Design Space Exploration in Robotics

Springer

Frank Ehlers
Kiel, Germany

ISSN 2198-4182 ISSN 2198-4190 (electronic)
Studies in Systems, Decision and Control
ISBN 978-3-031-78739-3 ISBN 978-3-031-78740-9 (eBook)
https://doi.org/10.1007/978-3-031-78740-9

This Springer imprint is published by the registered company Springer Nature Switzerland AG
The registered company address is: Gewerbestrasse 11, 6330 Cham, Switzerland

Love connects strings forever.

Preface

This book provides a foundation of the overall cycle from design and modelling to implementation and control of unmanned systems, exhibiting autonomy and instantiating self-organization amidst extreme disturbances, also functioning in uncertain and dynamic environments. The underlying assumption of the research path taken is that unmanned systems exhibiting attributes of autonomy, intelligence, autonomous functionality and resilience have to be considered as technically engineered systems. Hence, validation, verification, certification, and acceptance testing have to be provided together with the product "robot capable of fulfilling a specific requirement". The key is to provide a mathematically sound calculus to evaluate autonomy and autonomous functionality and resilience, directly applicable to real complex engineering systems. This overarching approach is reaching out to provide an ethically aligned view on engineering, specifically aiming to support human's free will. This support has two components: The first component is the automated code generation for the realisation of the idea the human has. The second component is the calculation of values that show how robust the newly invented apparatus can be, giving responsibility back to the human, now aware of potential deficiencies.

Readers are expected to be signal processing engineers and roboticists with interest in Digital Twins and Artificial Intelligence.

Many thanks to my students in the last years' lectures on Design Space Exploration for Unmanned Systems at the Christian-Albrechts-University Kiel who have been patient while the bigger picture connecting the building blocks from signal processing and system-of-systems design was developing. Many thanks to my colleagues who bear my theoretical view on practical problems to be solved. I am especially grateful to my family for providing the outer view on technically engineered systems.

Kiel, Germany
October 2024

Frank Ehlers

Contents

Chapter 1
Introduction

Abstract While science fiction often portrays robots with incredible capabilities and versatility, advancements in artificial intelligence, machine learning, and robotics are gradually enabling the development of sophisticated machines that can perform an array of tasks across diverse domains. These innovations have significant potential to revolutionize industries, enhance human capabilities, and shape our future society; however, it is essential to approach these developments with caution and ethical considerations in mind. The caution is on technical feasibility in terms of Verification, Validation, Certification, Acceptance Testing, and Trust. Together with ethical considerations, a design space is generated: the space of all parameter settings for all options a designer has in order to design a robot performing a specific task. In an engineering process, the specific task is formulated as a functional requirement, and the interesting work for a robot designer is to find an optimal operational setup within the design space to implement a robot performing the given task. If the given task is rather vaguely formulated, the robot needs to handle everything that is not explicitly preplanned, in an autonomous manner. Now, if the robot handles situations on its own, how can the designer then be sure that it handles everything correctly? What is the constraint for a robust and efficient autonomy in the design space? Let's explore together this area in the design space: Where is the robust and efficient autonomy region? How can we measure how far we are away from there in our current design? How can we formulate constraints in our design space such that the autonomy is working with sufficient robustness and efficiency?

This introduction has two parts: The first part is a story-telling way of where the problems are and which approaches are taken to tackle them. The second part is like a floor map of the different levels of abstraction for the reader. There, we understand Design Space Exploration as a process in which we walk through these different levels of abstraction.

F. Ehlers, *Design Space Exploration in Robotics*, Studies in Systems, Decision and Control 229, https://doi.org/10.1007/978-3-031-78740-9_1

1.1 A Toolset for Roboticists

Let's design a robot. How to find the right level of sophistication, the right level of details provided in models, the right sensor on the right robot? How to explain and certify the final choice? And how to accept robots having been built by others? To answer these questions the elements of the chapters in this Book dealing with Signals, Physics, Architectures, Robots, and Robot Coordination have to be combined; even more, the challenge of Design Space Exploration in Robotics is to describe and maintain the relationship between them in order to build the right robot, doing it right at the first time. The combination is possible if there is a foundation in a common task-related metric that is used in all related tools, i.e. risks and costs to miss a signal or to reach a goal have to be consistent with the formulation of the physical processes involved, with entries in architectures, with the programming of robots, and with software providing autonomously environmentally adaptive decision making on to be generated coordinated actuation signals. Of course, generating this glue-like metric comes with difficulties: The complexity of real-life experimentation needs to be described, and based on this description decisions on the relevancy of parameters and their interconnectivities have to be made. Then algorithms have to be written which are capable of such decision making in an online manner with limited processing power onboard. However, regarding the autonomous aspects, in the design phase different algorithmic approaches for such algorithms have to be evaluated: Which one is the best of these autonomy algorithms which are all reaching the given effective goal setting? Best, in the sense of "most efficient" or "most robust," or both? And shouldn't these autonomy metrics be task-agnostic, in the sense that the metrics describe the autonomy rather than repeating elements of the task?

This first part of the Introduction explains the research path taken in more details. We start at the signal processing task, link this to the physical constraints robots are operating in. Then, we concatenate the requirements of the robots' operation with a service-oriented architecture, which allows us to model the robots and their activities inside a common structure. The operationalization of this architectural structure allows us to address the Robot Coordination task analytically, with the help of introducing an implicit comparative metric, which we then contrast with an application of Human-Robot Collaboration in real-world scenarios: By taking this research path, we assume that we design a technically engineered system in which emergent features can only occur from already described elements, i.e. the autonomous system cannot explore details beyond a predefined smallest level of detailedness. Natural systems, instead, have access to make changes also below such smallest level, e.g. via the process of evolution. Also science is pushing the limits of detailed knowledge by utilizing natural intelligence, whereby both, evolution and natural intelligence, respectively, overcome the risk of getting stuck in local minima when utilizing comparative metrics.

Let us walk chapter by chapter through this Book: No matter in what kind of world (the fully complex nature world or a tiny example world) we let our robots operate in, the contact between the robots and the world is established via signals (see Chap. 2). We ask there how to interpret the incoming noisy signals, because in the real world every signal has an associated uncertainty.

When we continue to analyse signals from the robot's outer world, we can make the robot generating a model for the processes in this outer world (see Chap. 3). However, the robot will always have a limited precision on what is going on there, and even more, just like the robot receives only noisy signals, the outer world is driven by noisy signals. The outer world turns out to be quite large and describing all the sources, their interconnections, inter-dependencies, etc. needs some organisation, also if the robots need to share the information with others. But this organisation is not fixed, it depends on the application task: the operational goal which is a formulation of the reason why the robots are listening to all these signals, why they are collecting all the information. The robots need to reach the given operational goal, and they need to be careful over time not to run into difficulties about their ability to reach it, i.e. they have to control the situation.

Now that we can interpret signals and apply a control to get the robots reach the operational goal, we can step one level of abstraction upwards: We have seen that things match with services they are either part of or get manipulated by them. If the outer world has some structure, this thing-to-service abstraction seems to be a common feature. The idea is to describe the world and the robots by a service-oriented viewpoint, adding categories related to them and arrive at architectures (see Chap. 4). We introduce the Unified Architecture Framework (UAF), noting that there is a built-in path to combining architectures, letting the work with the UAF be described separately in a top-down and a bottom-up direction. For the top-down-direction: If a system is understood as a system that comprises of other systems, then the interaction between the sub-systems can be described with the help of the SysML, even one level more down with SysPhS as a link to the physical outer world. As a standard SysPhS provides entries into many physical thing and service descriptions, providing unified/standard names and units. For the bottom-up direction, an existing part (or building block) is interpreted as a little Enterprise: Incoming information is sold to a controller as customer, and the money earned is related to how much you cover the given operational goal. Each starting building block is considered to be a micro-service, and the next level of enterprise architecture is reached by combining the existing services with the help of a new business process. Where we end up is the digital description of a dynamic system, a digital twin, which ideally can be inserted into Dynamic Systems already described by the SysPhS in the top-down approach. Ideally does not mean automatically here. Hence, the confrontation of the two approaches (top-down vs. bottom-up) is certainly of an essential importance when exploring the design space.

We discuss in Chap. 5 how the languages developed in Architectures can be used to describe our (to be designed) robot in action, how the signals it receives can be used to generate a physical model of the environment it is operating in, leading to the finding that it should be possible to describe a Design Space Exploration-Machine (DSE-M), i.e. an algorithm that optimizes design and operational parameters for a given task. Putting the DSE-M into the robot would make the robot become an Artificially Intelligent Robot. This DSE-M is the start of the top-down approach, where strategies are given, the DSE-M is then selecting or generating references, and the references produce ontologies, which are loaded into a conceptual framework and compiled via generic programming approaches. As emergent features can be generated from these programms inside the robots, the bottom-up-performance calculations have to be used to link to the options from the top-down-approach efficiency and robustness, i.e. the real-time robot has to be modelled and this model has to become a part of the top-down-approach, and so on and so forth: We are describing a fixed-point algorithm here, and therefore we are looking for an analytical formulation, where it is possible to re-insert equations into equation syntactically. These equations, without any given structure to be exploited in computer-aided calculations will be very large, hence we restrict our approach to a formulation using polynoms with certain ordering features. Luckily, this restriction to polynomials offers a way to give a structure to the equations by re-formulating the forth-and-back in the fixed point algorithm in a recursive manner, starting with a tree-search at the lower degrees of the polynomial terms. In this Book, we are not discussing numerical implementations of the calculations, but refer here to the recent advances in parallel processing power that exists for artificial intelligence algorithm like deep neural networks: The formulas automatically generated by the DSE-M would be the chip-design and/or programming for such parallel processing power.

A preliminary for reaching this programming goal is to be able to set the three mathematical languages for service-based architectures, for noise signals and control in noise, and physical models in action all together to allow us to predict the performance of a robot on a specific task. This prediction is meant to be comparable to other robots, but not to other tasks. Such performance metric is the key to apply Artificial Intelligence algorithms in Robotics: Optimization algorithms can work on controlling to make the task easier and/or to make the performance of a specific robot better. The performance metric is not yet the autonomy metric, we are looking for in this Book. We need to separate the difficulty of the task from the capacity of the autonomous approach within the formulation of tasks' performance metrics. The autonomy metric is task-agnostic, hence it needs a generic approach to formulate a task performance metric, a metric for the difficulty of a task, and a metric for the autonomous approach. The methodology how to formulate and calculate the autonomy metric is the core contribution of this Book. The task performance metric for this methodology is restricted to scale-invariant formulations regarding the detailedness of the models to predict the task performance: The more we stay away from the critical value of the system being described by theses models, the more robust are the estimation of the task's difficulty and the autonomy metric.

In order to develop the generic approach, we will put in Chap. 6 our focus on the overlapping area of the two approaches (bottom-up, top-down) for the specific application of Multi-Robot Coordination. There, if we see a mismatch between the results of the two approaches, decisions have to be made with risks/costs following the performance metric we developed in Chap. 5. Influence Diagrams are a graphical way to describe this decision making. The values for the Influence Diagram are generated by the so-called Validation and Verification (V&V) process. Whilst Validation and Verification are representing the top-down and bottom-up directions, respectively, in a situation where bottom-up and top-down approaches match without inconsistencies, it cannot be forgotten that the system has to be implemented, i.e. becomes part of a real system with real noise, real hardware failures, real timing, etc. The Certification process, as an outer view on the designed system, continues the system analysis with the assumption that inconsistencies occurring inside the Field Robot (which might originally be originated also by non-resolved, but in the light of the given task acceptable inconsistencies between bottom-up and top-down approaches) are belonging to the discrepancy between measurement interpretation and reality, i.e. by the fact that within a Field Robot only an estimate of the a-priori probability can be extracted from data. For a technically engineered system then, we add a "final checking device", which is equipped with model information and data, to perform Certification as a final check, e.g. via formal models regarding externally and prior formulated certification rules. Coming to the final acceptance test, we need to apply a no-trust methodology and generate statistically significant data to make a final decision on whether the entire system is fulfilling its requirements, whereby no-trust means that we can only trust the measurements in the real world, not models nor simulations.

Contrasting (in Chap. 7) the view on ourselves as humans equipped with a free will (or as non-technically engineered objects) and the view on the Field Robots with an only as-good-as-possible level of top-down/bottom-up consistency, we as humans seem to be standing outside the processes listed in Chap. 6, leaving the work to be done there in principle to computers and their artificial intelligence. But is this purely technical view physically and ethically viable? As computations become expensive for complex systems, there will be a maximally affordable degree of resolution with an associated tolerance of system applications without success. Consciously understanding how to handle (including the option to avoid) decision-making is the next (or in a philosophical sense: the ever-lasting) big challenge of humanity.

1.2 Design Space Exploration as a Process

Despite the straight-forward story through the chapters of this Book told in the first part of this Introduction solving Design Space Exploration in Robotics is by far not straight-forward if only physical state variables are to be used. An impression of the necessity to tackle the Design Space Exploration task from different levels of abstraction is given with the link to the last chapter: It raises important questions, related to the fact that from an outside view the bottom-up path in the previous Section (see Fig. 1.1) has to be contrasted with a top-down path as in Fig. 1.2.

The different levels of abstraction can be mapped on the terms Machine, Robot, and Field Robot:

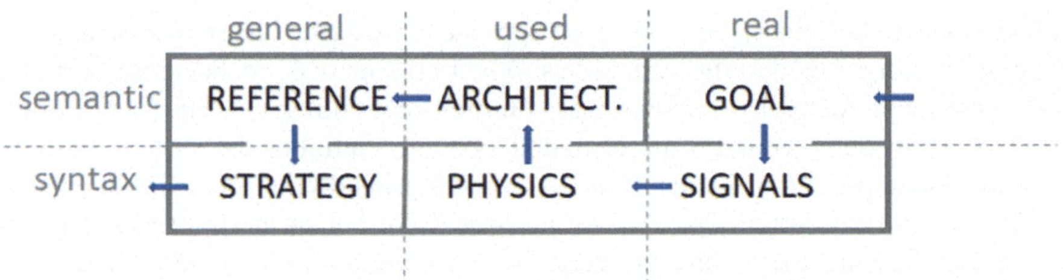

Fig. 1.1 Three different abstraction levels are needed to describe the Design Space Exploration: Coming from the real implementation, there is furthermore the used level, followed by a general level. Each of the levels has a subdivision in semantic and syntactical entries. In the story-telling way of the first part of this Introduction, we start with a design goal, imagine which sensors and actuators might be involved, describe the physics between them and between them and the given design goal, then we collect all these entries and relationships in an architecture. Looking at different architectures we identify patterns of design goal to design solution relationships, and finally we ask ourselves whether there might be a syntactical description of a strategy which we can use to associate different references with

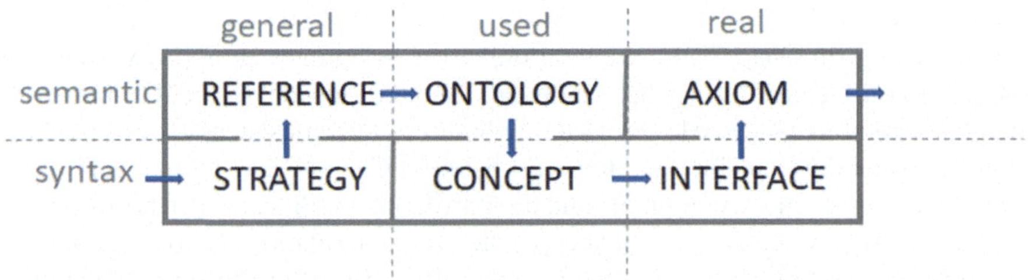

Fig. 1.2 If we have strategies to choose or generate references which generate an ontology which then can be conceptionally exploited to generate program code to deal with the signals from and to the interfaces to the real world, we can use these strategies to automatically design a Field Robot that reaches the design goal (marked as semantically true in the real world) and we can compare different outcomes of a strategy or the outcomes of different strategies

- Machine as hardware which is just executing fixed programming for a specific situation,
- Robot as a whiteboxing capability of passing through a given ontology to solve given tasks with awareness of the situation,
- Field Robot as a robot that can encounter emergence originated from within the technical models by autonomy as the capability of Implicit Ontology Generation on Demand.

In this Book, we investigate how to build a Field Robot.

This means, we insert (from a top-down view) into the ontology a robot that is prepared to handle all situations that can occur within the physical frame of maximum and minimum ranges of relevant descriptions. As already explained in the first part of this Introduction, we assume in this approach that for the first detailedness level we can find a solution, then checking all options for the next detailedness level, assuming that for each option we can find a solution, and so on and so forth. The resulting formulas describe then a cost function, where the same formula elements (we call them references) can occur when calculating the optima of this cost function, especially if we are using the same strategy for performing the optimization process (which is the process where we make the design decisions) for various tasks in various domains: On the one end of references we might find a swarm, on the other end a completely leveled hierarchy as best design decision for a given task in a given domain.

The Field Robot is equipped with a preset strategy which is used to select (or generate) a reference, which creates an ontology based on which the Field Robot can apply a Model-Based Code-Generation Concept to generate the code for dealing with the signals and actuators in such way that the Field Robot reaches its preset goal efficiently.

The Reader will be reminded at the beginning of each chapter on three different views:

- In the Abstract the part of Floor-Plan will be described in which the present chapter is located. For this part of the chapter start, we have to maintain our view on Design Space Exploration as the task to analytically combine a syntactic strategy formulation with a semantic axiom truth to be reached in the real world.
- In the Top-Down View, autonomy is viewed as a concept.
- In the Story-Telling path of the first part of this Introduction the Bottom-Up View on the Field Robot is described.

We will link the strategy-to-axiom connection with the receivers-operating-characteristics, whereby manipulation tasks are linked to a predicted receivers-operating characteristics (a projection on what the field robot will be able to achieve) and we will look for connections that allow a gentle degradation of descriptive performance when lowering the level of detailledness of the description.

We will link the autonomy concept to the autonomy metric: How well does a specific autonomy implementation will work for a specific task or specific category of tasks, which is described independently from the difficulty of the tasks. The gentle degredation of the descriptive performance is a preliminary to derive an autonomy metric and an evaluation of the difficulty of a specific task or specific category of tasks.

At the end of each chapter, a paragraph marked as Take-Away for Design Space Exploration will help the Reader to develop a holistic view on Design Space Exploration in Robotics. At the end of the day, Design Space Exploration is the formulation of an optimization process which results in Design Decision. But the optimization process has to come to a solution, which has to be robust and efficient, opening the door for follow-on certification and acceptance processes. Design Space Exploration can certainly be performed differently from the approach presented in this Book. The Take-Away for Design Space Exploration gives a rational behind this Book's approach.

Remark

As the Reader can see from the Table of Contents, the number of subjects covered is high and, furthermore, subjects are stemming from different scientific/engineering domains. The objective of this Book is NOT to give a comprehensive introduction into all of these subjects. Therefore, in a minimalistic manner only a few entry points into each of the various subjects are given in the references sections. The recommendation for the interested Reader is to check these entry points, and with the keywords in the associated sections, the large amount of online available (and constantly growing) material.

So, what is the objective of this Book then? The objective is to form from these subjects a consistent/holistic tool chain for Design Space Exploration in Robotics. Even more, the objective is to generically allow for a mathematically sound transition between the two views: the detailed inside design view and the outside view on the Field Robot's behavior. Just like in physical experiment design, the task in Design Space Exploration is to model the microscopic-to-macroscopic transition and to exploit the knowledge about this transition in mathematical optimization processes. The better we organize this transition, the easier we can we can setup a trustworthy (because of mathematical proofs available behind the scene) tool chain for the design of Field Robots. In the highest abstraction level, the book wants to point to one strategy on how to setup a tool chain for the design of Field Robots with the presented subjects as abstract references.

Within these presented subjects/references, the Blind Separation solutions turned out to be less well-represented in the literature or in online available sources. Hence, for the Blind Separation also a comprehensive introduction is given.

Chapter 2
Signals

Abstract In this Chapter, we imagine the Field Robot as an observer of the signals received from its sensors. As the signals are corrupted by noise, uncertainty about the meaning and the origin of the signals needs to be captured within the algorithms that generate actuations given the input sensor signals. There are different ways to describe and process this uncertainty, and in the first part of this chapter we list the frequentist's, the Bayesian, and Dempster-Shafer's approaches. In the second part of this chapter, we look at ways how to combine the three different formulations. We introduce the FISST, BGoF and Blind Separation methods. In the third part, we apply the three different ways to Decision-Making tasks, finding that the three approaches fit to different settings: frequentists for fusion with no prior information, Bayesian for fusion with prior information, Dempster-Shafer for fusion with inconsistencies. The Field Robot will need a mixture of these views: as less data as possible, as less prior information as possible, as less inconsistencies as possible. Hence, the best Field Robot Design has to apply a sequence/mixture of FISST, BGoF, and Blind Separation on the incoming data to minimize the amount of data (resulting in quick reaction), the amount of pre-programming (resulting in higher autonomy), and the value associated to inconsistency (resulting in higher robustness).

Top-Down View
If a signal comes in as a sample what does this sample tell us about the semantic truths related to it? For example, if a light sensor sends a binary 1010, can we infer that the sun is shining or not? As a strategy, we can collect data and correlate it with ground truth. As a concept, we can model the physics of light emission and light absorbing devices, generate a data-to-state relationship and assign a value that relates to the uncertainty that the binary 1010 refers to the shining sun. Or, as a third possibility, we can exclude all other light sources, establishing a non-noisy information transmission paths that tells us (in an aximomatic manner) the binary value of a light sensor in relationship to photons the sun emits. As Fig. 2.1 depicts, we have inserted three relationships to the sample in the interface.

F. Ehlers, *Design Space Exploration in Robotics*, Studies in Systems, Decision and Control 229, https://doi.org/10.1007/978-3-031-78740-9_2

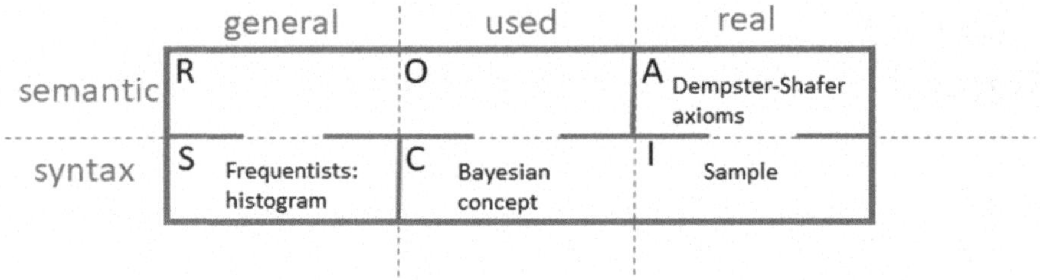

Fig. 2.1 The Field Robot receives samples. Semantically there is an axiomatic truth originating this sample. The Bayesian concept is capable to describe (syntactically) the knowledge the Field Robot can generate from this sample. However, quantitatively this descriptions depends on a prior probability, which has to be consistent with the measurement. Many samples can generally be stored in histogram, but histogram parameters have to be chosen, like number of bins, size of bins, handling of outliers, etc.

From the first part of the Introduction, we repeat that the relationship between a robot and the outer world is established via the signals received through the interfaces. The sensors of the robot send signals to its processing unit, where also signals from the inner processes of the robot are to be fused in order to get a consistent model of the current situation.

2.1 Three Ways to Look at Probability

Signals contain mixtures of different information sources, signals can be corrupted, signals can be affected by additional noise sources, signals are sent via channels that cannot carry all information provided by an information source, and an information source does not provide all information about the state of the object that sends this information.

In this Section, we look from three different perspectives of what is described by the concept of probability: frequentist statistics, Bayes concept, and the Dempster-Shafer theory.

Frequentist statistics, a classical approach to probability theory, relies on the concept of repeated experiments or observations to determine probabilities by counting the frequency of occurrence for specific outcomes. This method assumes that long-run relative frequencies will converge towards fixed and well-defined probabilities. While it has been widely used in various fields such as statistics, economics, and finance, Bayesian methods have gained prominence due to their ability to incorporate prior knowledge and update beliefs based on new data, providing a more flexible and adaptive approach for decision-making under uncertainty. The Dempster-Shafer theory (DST) offers an alternative framework to classical probability theories, allowing for the representation and manipulation of uncertainty in a more general setting. Unlike traditional probabilities that assign single values to events, DST assigns belief masses over sets of mutually exclusive hypotheses. This enables the integration of imprecise or conflicting information while preserving coherence between different sources of evidence. The theory has found applications in fields such as decision-making under uncertainty, data fusion, and expert opinion aggregation, providing a more robust approach to handling ambiguity and lack of information.

2.1.1 Frequentist View

When repeating measurements, the stunning observation is that although the same measurement equipment is used the measured values vary. Scatter plots and histograms, among a lot of other display tools of sometimes enormous sophistication, visualize this effect. Measuring a height of a tree 100 times will lead to a histogram of values around the true height. The measurement process has to be described in its very details: Is the histogram displaying 100 times measurements with the same equipment? Or is it measured once with 100 measurement equipments? Different effects might occur for different setups. And finally, after the measurement has been made, the analysis tells us: 90% of all measurements show values between x and y, etc. When we measure the next time, we can interpret these values as a probability for the value that we measure. In the next subsection, we use a probabilty value to describe a degree of belief, which is a different starting point than the frequentists data-driven approach.

2.1.2 Bayesian View

Assume we have some prior knowledge about the shape of the histogram generated by the frequentist's approach in the previous subsection. Then we can use this prior knowledge to infer from a set of data gathered the state of the system producing this data. This inference works in terms of classification (i.e. grouping the data) and in terms of data fusion (sensor-wise and/or time-wise).

We are giving in the next two paragraphs two example application of the Bayesian view, in the light of its usage inside a Field Robot. Various algorithmic approaches to Bayesian algorithms are possible, because working on practical problems with a limited amount of data is a challenging task. In this Book we will stay in the analytical realm, assuming also the availability of an almost infinity amount of data, as will be explained later, hence the introduction of the Bayesian view in its direct simple form is already sufficient for the purpose of utilizing the Bayesian view in this Book.

2.1.2.1 Bayesian Classification with Mixture Model

Following the approach in [9], we first consider the classification problem with the goal of predicting an unordered categorical response viable $y \in \{1, \ldots, C\}$ from a vector of real-valued predictors $x = (x_1, \ldots, x_p)$. That is, we want to give the Field Robot the ability to organize the data it is receiving. One could think about performing this classification task before building a robot. This is of course possible for very stationary environments, but with the Field Robot design we are aiming at a robot that is capable to handle instationarily occurring new situations. Many options arise at this point to mix approaches: complete prior knowledge being delivered as a static program element to the Field Robot, learning examples with a supervised classification being delivered to the Field Robot, or at the other end of such continuum of possibilities the implementation of unsupervised learning technics to order the data. Here, in this Section, we deal with a fully supervised or semi-supervised classification. Later in this Chapter, we introduce a method to separate data completely without supervision (a technique called Blind Separation). In fully supervised classification problems, both y_i and x_i are observed for every item in a training sample of n items, while in semi-supervised classification x_i is observed for $i = 1, \ldots, n$, but y_i is observed only for a subset of $n_0 < n$ of labeled items. Instead of modeling the class probability directly, we can model the marginal likelihood of y_i and the conditional likelihood of x_i given y_i to induce a model on the classification function using Bayes' rule:

$$Pr(y_i = c|x_i) = \frac{Pr(y_i = c) f(x_i|y_i = c)}{\sum_{c'=1}^{C} Pr(y_i = c') f(x_i|y_i = c')} \tag{2.1}$$

In Bayesian discriminant analysis, we specify a prior for the marginal category probabilities $\psi = (\psi_1, \ldots, \psi_C)$, where $\psi_c = Pr(y_i = c)$, and for the collection of category-specific conditional likelihoods $\{f_c\}_{c=1}^{C}$. A simple conjugate prior for ψ is the Dirichlet prior. Suppose we want to model the joint density of x_i flexibly, where x_i' comprises a mixture of continuous and categorical predictors. Let $f(x|\theta_i) = \Pi_{j=1}^{P} K_j(x_{ij}|\theta_{ij})$ with K_j a kernel appropriate to the support of the jth predictor. This product kernel model assumes conditional independence given θ_i. To induce dependence, we again rely on the mixture idea to assign a joint prior distribution on the full complement of θ_i,

$$\theta_i \tilde{P} = \sum_{h=1}^{H} \pi_h \delta_{\Theta_h}, \Theta_h = (\Theta_{h1}, \ldots, \Theta_{hp}). \tag{2.2}$$

where δ_Θ denotes a degenerate distribution with all its mass concentrated at Θ so that $\theta_i = \Theta_h$ with probability ψ_h. We let $\Theta_{hj} \tilde{P}_{0j}$ independently, with P_{0j} a prior appropriate of the jth kernel. As some simple examples, we may consider Gaussian kernels for continuous predictors, Bernoulli kernels for binary predictors and Poisson kernels for count predictors, with the corresponding P_{0j}'s chosen to be conjugate.

The above product kernel mixture is also a simple latent class model. Let $z_i = h$ denote the latent class for item i, with $Pr(S_i = h) = \pi_h$. Individuals in latent class (subpopulation) h have distinct parameters within each kernel. Elements of x_i are conditionally independent given z_i, but are marginally dependent. We can use the product kernel approach in the classification setting to jointly model the distribution of y_i and x_i.

$$Pr(y_i = c|z_i = h) = \psi_{z_i c},\ f(x_i|z_i = h) = N(x_i|\mu_{z_i}, \sigma_{z_i}) \tag{2.3}$$

Averaging over z_i, we induce

$$Pr(y_i = c|x_i) = \frac{\sum_{h=1}^{H} \pi_n \psi_{hc} N(x_i|\mu_h, \sigma_h)}{\sum_{h=1}^{H} \pi_n N(x_i|\mu_h, \sigma_h)}, \tag{2.4}$$

which can be shown to be extremely flexible. We can also do the same for the elements of the x_i; for example, specifying separate multinomials for categorical predicturs and normals for continuous predictors.

Two important ideas arise from this classification approach: What if a robot would be available to further investigate the latent subpopulation h? What if the representation of the degenerate distribution δ_Θ is approximated by a Gaussian which is then represented by a polynomial with an increasing degree? Can we incorporate a cascade of robots successively investigating (always for an higher degree of polynomial description) the subpopulations?

2.1.2.2 Bayesian Tracking and Data Fusion

Following [6], we want to elaborate also on a time-wise application of the Bayesian concept for nonlinear tracking. Tracking means the concatenation of contact information in a way that it can be associated to the movement of a target. The Field Robot is either given some model knowledge about the system it is observing, or it can even be capable of deriving/estimating this system knowledge during its deployment. Tracking can be important for both, internal and external state variables, e.g. the tracking of an object hold by a Field Robot's own manipulator. Tracking can be done so-to-say in two time directions: forward and backward (retrospective) on the data already available, but also in a predictive/simulating manner on how the states variables will evolve in the future. This is important for the Field Robot's control and path-planning.

We consider the evolution of the state sequence $\{x_k, k \in N\}$ by a so called system model:

$$x_k = f_k(x_{k-1}, v_{k-1}) \tag{2.5}$$

where $f_k : R^{n_x} \times R^{n_v} \to R^{n_x}$ is a possible nonlinear function of the state x_{k-1}, $\{v_{k-1}, k \in N\}$ is an i.i.d. process noise sequence, n_x, n_v are dimensions of the state and process noise vectors, respectively, and N is the set of natural numbers. The objective of tracking is to recursively estimate x_k values from measurements via the so called measurement model:

$$z_k = h_k(x_k, n_{,k}) \tag{2.6}$$

where $h_k : R^{n_z} \times R^{n_n} \rightarrow R^{n_z}$ is a possible nonlinear function, $\{n_k, k \in N\}$ is an i.i.d. measurement noise sequence, and n_z, n_n are dimensions of the measurement and measurement noise vectors, respectively. In particular, we seek filtering estimates of x_k based on the set of all available measurements $z_{1:k} = \{z_i, i = 1, \ldots, k\}$ up to time k.

From a Bayesian perspective, the tracking problem is to recursively calculate some degree of belief in the state x_k at time k, taking different values, given the data $z_{1:k}$ up to time k. Thus, it is required to construct the probability density function (pdf) $p(x_k|z_{1:k})$. It is assumed that the initial pdf $p(x_0|z_0) = p(x_0)$ of the state vector, which is also known as the prior, is available (z_0 being the set of no measurements). In the Sect. 2.2 of this chapter we will further elaborate on this assumption because it is essential for a robot to work with priors that reflect the reality. Continuing with the nonlinear tracking, the pdf $p(x_k|z_{1:k})$ is recursively calculated in two stages: prediction and update.

Suppose that the required pdf $p(x_{k-1}|z_{1:k-1})$ at time $k-1$ is available. The prediction stage involves using the system model to obtain the prior pdf of the state at time k via the Chapman-Kolmogorov equation

$$p(x_k|z_{1:k-1}) = \int p(x_k|x_{k-1})p(x_{k-1}|z_{1:k-1})dx_{k-1}. \tag{2.7}$$

A key point is to use $p(x_k|x_{k-1}, z_{1:k-1}) = p(x_k|x_{k-1})$ as the system model describes a Markov process of order one. The probabilistic model of the state evolution $p(x_k|x_{k-1})$ is defined by the system equation and the known statistics of v_{k-1}.

At time k, a measurement z_k becomes available, and leads to updating the prior via Bayes' rule

$$p(x_k|z_{1:k}) = \frac{p(z_k|x_k)p(x_k|z_{1:k-1})}{p(z_k|z_{1:k-1})} \tag{2.8}$$

where the normalizing constant

$$p(z_k|z_{1:k-1}) = \int p(z_k|x_k)p(x_k|z_{1:k-1}dx_k \tag{2.9}$$

depends on the likelihood function $p(z_k|x_k)$ defined by the measurement model and the known statistics of n_k. In the update stage, the measurement z_k is used to modify the prior density to obtain the required posterior density of the current state.

The recurrence relation can also be written down as a (probably long) formula as an entry in a Design Space Exploration task as generic sensor filter. In this Book, we will aim at arriving at a system and measurement model with only Gaussian noise sources, opening the possibility of maintaining an generic (because analytic) treatment throughout the entire Design Space Exploration process.

2.1.3 Dempster-Shafer Theory

Following [3], the *Belief Theory Approach* (or also called Dempster-Shafer Theory, which is the name that is used later in this Book in order not to confuse the word belief in the Bayesian and also other mathematical frameworks in later Chapters) allows the interpretation of statements with a probabilistic formulation. It is a mathematically approach to combine such statements. For the Field Robot this theory has several immediate applications: It can be used to quantitatively rate inconsistencies, e.g. between plans and actual estimation for state variable or between information coming in through external communication channels.

2.1.3.1 Combination of Beliefs

- Dempster-Shafer start with an exhaustive (i.e. complete) set of mutually exclusive outcomes of some experiment of interest, S, and call it the "frame of discernment".
- They form $\Omega = 2^S$, and assign belief $B(A)$ to any set $A \subset \Omega$.
- Example: $S = \{x, y, z\}$, $\Omega = \{\emptyset, \{x\}, \{y\}, \{z\}, \{x, y\}, \{x, z\}, \{y, z\}, S\}$
- The elements of S can be called atomic events; the elements of Ω can be called molecular if they are not atomic.
- The interpretation of a molecular event is that any one of its atomic elements is "in it", but not in a constructive sense.
- The evidence assigned to a molecular event cannot be titrated; it applies to the molecular event as a whole.
- Basic probability assignment (or mass of evidence)
  It satisfies the following axioms:
  - $m(\emptyset) = 0$
  - $m(A) > 0 \quad \forall \text{ events } \; A \epsilon \Omega$
  - $\sum_{A \subseteq \Omega}(A) = 1$
- Note: These axioms look similar to the axioms for probability, **but** they are **not** the same. Belief and probability are not identical. Crucial difference is that axiom 3 equates unity with the total accumulated evidence assigned to all elements of Ω, whereas an axiom of probability equates unity with $S \in \Omega$.
- Belief theorists interpret S to mean a state of maximal ignorance. Hence, in the absence of any evidence, in a state of total ignorance, assign $m(S) = 1$.

- If knowledge increases, then $m(A) > 0$ for some $A \epsilon \Omega$, and $m(S) < 1$ in accordance with the reduction of ignorance.
- "Ignorance" in belief theory is a very useful property of the approach, but it is not unique to belief theory.
- Belief theory further defines a belief function in terms of the mass of evidence. The mass of evidence assigned to a particular set is committed exactly to the set, and not to any of the constituent elements of the set.
  Therefore, to obtain a measurement of total belief committed to the set, add the masses of evidence associated with all sets that are subsets of the given set. For all sets A and B in Ω, define the belief function

$$\mathrm{Bel}(A) = \sum_{B \subseteq A} m(B) \tag{2.10}$$

- $$m(B) = \sum_{A \subseteq B} (-1)^{|B|-|A|} \mathrm{Bel}(A) \tag{2.11}$$

- $\mathrm{Bel}(A) = \mathrm{Cr}(A) = \mathrm{Su}(A)$
  Belief is sometimes called Credibility or Support.
- Plausibility

$$\mathrm{Pl}(A) = 1 - \mathrm{Bel}(\mathrm{not}A) = 1 - \sum_{A \cap B = \emptyset} m(B) = \sum_{A \cap B \neq \emptyset} m(B) \tag{2.12}$$

- For any set $A \in \Omega$:

$$\begin{aligned} \mathrm{Su}(A) &\leq \mathrm{Pl}(A) \\ \mathrm{Su}(A) + \mathrm{Su}(\mathrm{not}A) &\leq 1 \\ \mathrm{Pl}(A) + \mathrm{Pl}(\mathrm{not}A) &\geq 1 \end{aligned}$$

- "Support" means "belief already available".
- "Plausibility" means "belief that may become available".

Suppose $S = \{x, y, z\}$, $\Omega = \{\emptyset, \{x\}, \{y\}, \{z\}, \{x, y\}, \{x, z\}, \{y, z\}, S\}$ (Table 2.1).

From the point of view to have the task to program a robot, the best situation would be if information for all atomic elements were available. This situation would allow to continue with e.g. the Bayesian concept, filtering out the information necessary for the task and progress with optimal control along an optimal planning.

For the Design Space Exploration we see the robot to receive data which can be put into histograms, if the robot would already knows the categories behind the histogram, which in the case of a Field Robot in an emergent environment are not even present at design time. Hence, to continue our Design Space Exploration, now we have to find a generic way, to combine samples in a unconstructed environment, extract information from the available samples and predict future samples. In the next

Table 2.1 Example for the application of Dempster-Shafer theory

Event A	$m(A)$	$\mathrm{Su}(A)$	$\mathrm{Pl}(A)$
$\emptyset$	0	0	0
$\{x\}$	m_x	m_x	$1 - m_y - m_z - m_{yz}$
$\{y\}$	m_y	m_y	$1 - m_x - m_z - m_{xz}$
$\{y\}$	m_y	m_y	$1 - m_x - m_z - m_{xz}$
$\{xy\}$	m_{xy}	$m_x + m_y + m_{xy}$	$1 - m_z$
$\{xz\}$	m_{xz}	$m_x + m_z + m_{xz}$	$1 - m_y$
$\{yz\}$	m_{yz}	$m_y + m_z + m_{yz}$	$1 - m_x$
S	$1 - \sum$ all other masses	1	1

section, we see how we can step forward with this plan by combining the different view-points on probability.

2.2 Three Ways to Combine Different Views on Probability

In this Section, we elaborate on the idea that Finite Set methods (FISST) can combine the Dempster-Shafer theory and the Bayes concept. We repeat elements of the Bayesian Goodness-of-Fit algorithm which in an iterative generic manner combines the Bayesian concept with the frequentist strategy. Furthermore, we find that Blind Source Separation can be a between the frequentist strategy and the access to atomic elements in the Dempster-Shafer theory.

2.2.1 FISST

Following [2], the finite set method is an advanced technique for estimating the state, trajectory, and number of targets in a multi-target system. It offers significant improvements over traditional methods by explicitly handling data association ambiguities and allowing for more accurate tracking even when faced with missing, false, or cluttered information. This approach has been widely adopted due to its ability to efficiently handle complex scenarios encountered in modern surveillance and defense systems, leading to improved performance and decision-making capabilities.

We repeat from the elaborations of three interesting features of FISST that seem to be essential elements when we look at a Field Robot handling the data it receives through its sensors. To be generic, we need a mathematical formulation for the imprecise, contingent, and ambiguous data problems. We cannot go into the details of the implementation of FISST-based algorithms here (e.g. [4]), but we need to give the Reader an impression of the potential FISST-based algorithms have when applied to scenarios involving Field Robots. Since FISST combines the structure of Bayesian formulations with imprecise, contingent and ambiguous data, it fits to scenarios we foresee Field Robots find themselves in.

2.2.1.1 Imprecise Data: Dempster-Shafer Bodies of Evidence

A Dempster-Shafer body of evidence B on some space U consists of nonempty subsets $B : B_1, \ldots, B_b$ of U and nonnegative weights $b_1, \ldots, b_b$ that sum to one. The Dempster-Shafer theory can be generalized to the case when the B_i are fuzzy membership functions. Such fuzzy bodies of evidence can also be represented in random set form.

2.2.1.2 Contingent Data: Conditional Event Algebra

Knowledge-based rules have the form X follows S = if X then S where S, X are subsets of a (finite) universe U. There is at least one way to represent knowledge-based rules in random set form.

2.2.1.3 True Likelihood Functions for Ambiguous Data

The core step in a strict Bayesian formulation of the ambiguous data problem is to specify a likelihood function for ambiguous evidence that models the understood likelihood that a specific ambiguous datum Θ will be observed, given that a target of state x is present. This is where practical problems are encountered. The required likelihood function must have the form

$$f(\Theta|x) = \frac{Pr(R = \Theta, X = x)}{Pr(X = x)} \tag{2.13}$$

where R is a random variable that ranges over all random closed subsets Θ of the measurement space. However, $f(\Theta|x)$ cannot be a likelihood function unless it satisfies a normality equation of the form $\int f(\Theta|x)d\Theta = 1$ where $\int f(\Theta|x)d\Theta$ is an integral that sums over all closed random subsets of measurement space. No clear means exists for constructing a likelihood function $f(\Theta|x)$ that not only models a particular real-world situation, but also provably integrated to unity. If $f(\Theta|x)$ could be specified with sufficient exactitude, a high-fidelity conventional likelihood $f(z|x)$ could be constructed.

2.2.1.4 DSE Perspective on FISST

The summary of this short look into finite set mathematics with the perspective on Design Space Exploration in Robotics leads to two preliminary heuristics:

- Be honest to your data processing tools, in the sense that there is a mathematical tool at hand which allows to formulate all doubt about models and data, and at the same time allows to insert all the prior knowledge about physics and processes related to models and data.
- Try to find out what sufficient exactitude means for your application. Clearly, in general, constructing a true likelihood function is not the aim, when doing Design Space Exploration for Robotics with given tasks for the robots. We have to link sufficiency to the task, whereby this linkage has to be generic (i.e. a mathematical formula).

Fig. 2.2 depicts that the power of FISST is to combine the Bayesian concept with the axioms in the Dempster-Shafer theory.

2.2.2 *Bayesian Goodness of Fit (BGoF)*

In this Subsection, we echo the statements about the Bayesian Goodness of Fit algorithm from its developers [5]. BGoF is a Bayesian data modeling scheme that performs four interconnected tasks:

1. characterizes the uncertainty of the elicited parametric prior;
2. provides exploratory diagnostic for checking prior-data conflict;
3. computes the final statistical prior density estimate; and
4. executes macro- and micro-inference.

In [5] a general modeling framework, along with concrete guidelines for applied users is given. They introduce a new class of nonparametric priors $DS(G, m)$ along with its role in exploratory graphical diagnostic and uncertainty quantification. The

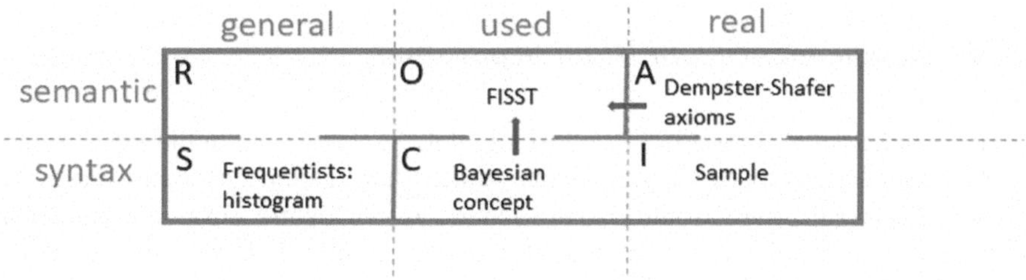

Fig. 2.2 If we utilize the power of FISST to fuse the Bayesian concept and the Dempster-Shafer axiomatic meaning, we are able to model the semantics of data analysis that has to happen in a Field Robot.

modified shifted Legendre polynomials are used in their model, with the model-building approach proceeding step-by-step through:

1. it starts with a scientific (or empirical) parametric prior $g(\theta; \alpha, \beta)$,
2. inspects the adequacy and the remaining uncertainty of the elicited prior using a graphical exploratory tool,
3. estimates the necessary correction for assumed g by looking at the data,
4. generates the final statistical estimate, and
5. executes macro and micro-level inference.

By that, they present a new family of prior densities which have a Fourier non-parametric representation in a specialized orthonormal basis. The model specification further read as:

- It has a unique two-component structure that combines assumed parametric g with the d-function. The function d can be viewed as a "correction" density to counter the possible misspecification bias of g.
- The density function $d(u; G, \Pi)$ can also be viewed as describing the "excess" uncertainty of the assumed $g(\theta; \alpha, \beta)$. For that reason we call it the U-function.
- The motivation behind the representation chosen in BGoF stems from the observation that $d[G(\theta); G, \Pi]$ is in fact the prior density-ratio $\pi(\theta)/g(\theta)$.

An iterative generic algorithm (called Methods of Moments) is presented which does not include numerical optimization routines, i.e. a lengthy polynomial expression can be written down to represent the estimation of the prior density, given data and model knowledge (in the Bayesian notation). This is an important finding for our approach to Design Space Exploration: We can model a Field Robot performing BGoF by directly inserting such (long) formula for data processing to samples that we simulate or describe by another analytic formula (as we will do in the next Chapter), and we can insert this formula for a Bayesian prior into Bayesian algorithms like classification, tracking or decision-making.

With BGoF the to be developed robot has the possibility to extract prior information from the histogram building with sampled data. The frequentist strategy is now linked to Bayesian concept which FISST links to the Dempster-Shafer theory.

2.2.3 Independence and Blind Separation

After the discussion of the FISST and BGoF algorithms, we see in the Fig. 2.3 that a direct connectivity between frequentist strategy and Dempster-Shafer theory is missing. This connectivity would lead to a generic way of identifying atomic elements in Dempster-Shafer, that can then be treated as (sufficiently) independent sources of information. The phrase sufficiently refers to the fact that as the Bayes recursive formular and BGoF Method of Moments also the Blind Separation can be treated as an iterative algorithm, which can be used to build (long) formulas with polynomic entries. If we see a cycle of data coming, extract independent source, estimate model

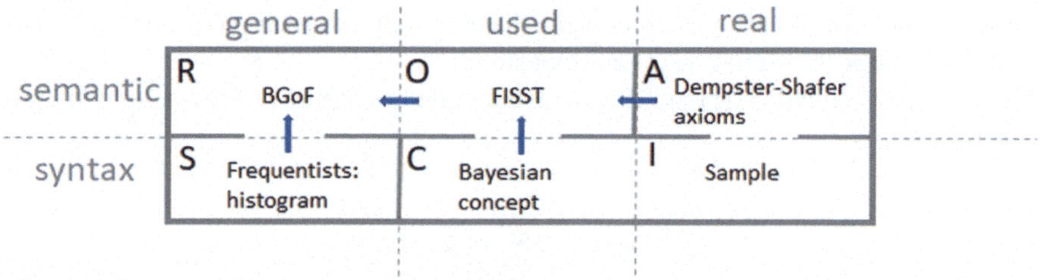

Fig. 2.3 If we utilize the power of BGoF to fuse the Bayesian concept and frequentist's histogram development, we are able to generally derive prior probability settings from data.

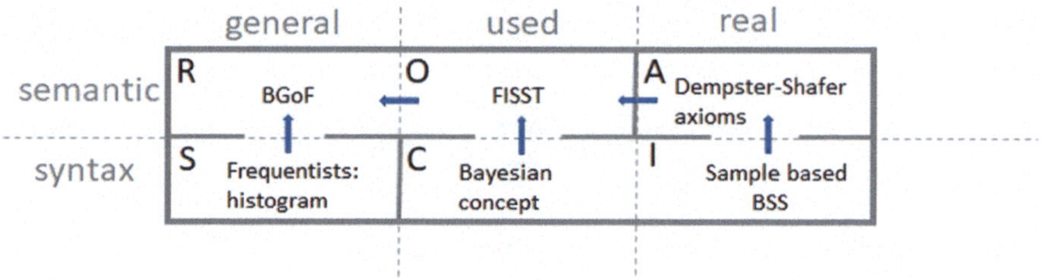

Fig. 2.4 The Blind Separation technique can calculate from frequentist's sample moments separated atomic axioms in the Dempster-Shafer sense.

parameters, use the model developed by the previous two steps to make predictions on what might happen next given some possible action, then we can further think about the robot as checking all possible actions and choosing the best action to reach the goal of the given task. In other words, blindly separating the data has the ability to enable the generic modelling of a Field Robot programmed in the methodology of behavioral robotics.

In this Subsection, we introduce the Blind Separation technique in more details, because in the following Chapters we will re-use this technique in two other abstraction-modes. Hence, it is necessary to understand the assumptions and constraints of the Blind Separation technique. We start the introduction of Blind Separation with three levels of difficulty: linear instantaneous, linear with time delay, nonlinear.

2.2.3.1 Blind Separation, Linear Instantaneous

The simplest task in signal separation without a-priori information about the source signals and the mixing coefficients occurs if N sources are mixed in a linear and instantaneous way. The mathematical formulation for this task is:

1. the existence of N source processes, that transmit the signals $a_1(t), a_2(t), \ldots, a_N(t)$. All signal values $a_j(t)$ are unknown for all time values t.
2. the source processes are independent, i.e. independent source signals

3. the existence of M $(M \geq N)$ receivers, that receive the signals $i_1(t), i_2(t), \ldots, i_M(t)$. The values of the signals $i_k(t)$ are measured.
4. the received signals $i_k(t)$ are linear instantaneous mixtures of the source signals $a_j(t)$, i.e.:

$$
\begin{aligned}
i_1(t) &= \sum_{j=1}^{N} c_{1j} a_j(t); \\
i_2(t) &= \sum_{j=1}^{N} c_{2j} a_j(t); \\
\vdots \quad & \quad \vdots \\
i_M(t) &= \sum_{j=1}^{N} c_{Mj} a_j(t),
\end{aligned}
\tag{2.14}
$$

where the mixing coefficients c_{jk} are a-priori unknown.

By calculating a sufficient number of (time-delayed) correlations and/or higher order (time-delayed) cumulants of the mixture signals, it is possible to estimate the mixing coefficients only with analysis of the measured values $i_k(t)$ and invert the mixture with the mixing coefficients, i.e., estimate the values of the source signals.

2.2.3.2 Blind Separation, Linear with Time Delay

Given are the inputs

$$
\begin{aligned}
i_1(t) =& a_1(t) + c_{12} a_2(t - \vartheta_{12}) \\
i_2(t) =& a_2(t) + c_{21} a_1(t - \vartheta_{21})
\end{aligned}
$$

The delays $\vartheta_{ij} \quad i = 1, 2;\ j = 1, 2$ are given.

The independent source signals $a_1(t), a_2(t)$ and the mixing coefficients c_{12}, c_{21} are unknown.

We describe how a Neural Network can calculate outputs

$$
\begin{aligned}
u_1(t) =& i_1(t) + c_{12} i_2(t - \vartheta_{12}) \\
u_2(t) =& i_2(t) + c_{21} i_1(t - \vartheta_{21})
\end{aligned}
$$

that separate the mixtures of the source signals.

The mixing coefficients c_{12}, c_{21} are unknown. So the algorithm has to start with random de-mixing coefficients, say t_{12}, t_{21}.

The output signals of the Neural Network are calculated:

$$\begin{aligned} u_1(t) =& i_1(t) - t_{12} i_2(t - \vartheta_{12}) \\ =& a_1(t) + c_{12} a_2(t - \vartheta_{12}) - t_{12}(c_{21} a_1(t - \vartheta_{12} - \vartheta_{21}) \\ & + a_2(t - \vartheta_{12})) \\ =& a_1(t) + (-t_{12} c_{21} a_1(t - \vartheta_{12} - \vartheta_{21})) \\ & + a_2(t - \vartheta_{12})\,(c_{12} - t_{12}) \end{aligned} \tag{2.15}$$

$$\begin{aligned} u_2(t) =& -t_{21} i_1(t - \vartheta_{21}) + i_2(t) \\ =& -t_{21} a_1(t - \vartheta_{21}) - t_{21} c_{12} a_2(t - \vartheta_{12} - \vartheta_{21}) \\ & + c_{21} a_1(t - \vartheta_{21}) + a_2(t) \\ =& a_1(t - \vartheta_{21})\,(c_{21} - t_{21}) \\ & + a_2(t) - t_{21} c_{12} a_2(t - \vartheta_{12} - \vartheta_{21}) \end{aligned} \tag{2.16}$$

If

$$t_{12} = c_{12}; \qquad t_{21} = c_{21}$$

the output of the neurons are

$$u_1(t) = a_1(t) - c_{12} c_{21} a_1(t - \vartheta_{12} - \vartheta_{21}) \tag{2.17}$$

$$u_2(t) = a_2(t) - c_{12} c_{21} a_2(t - \vartheta_{12} - \vartheta_{21}) \tag{2.18}$$

To find these values of t_{12}, t_{21} the gradient descent algorithm is used:

$$\dot{t}_{ij} = -\epsilon \frac{\partial V}{\partial t_{ij}} \tag{2.19}$$

where V is a cost function, i.e. V has a global minimum for the correct de-mixing coefficients.

A possible cost function is

$$V(T) = \sum_{i \neq j} \langle u_i(t) u_j(t) \rangle^2 + \langle u_i(t) u_j(t + \tau_0) \rangle^2 \tag{2.20}$$

Then, the update steps of the gradient descent algorithms are

$$\begin{aligned} \frac{\partial V}{\partial t_{12}} =& 4\langle (-i_2(t - \vartheta_{12}))\, u_2(t) \rangle \langle u_1(t) u_2(t) \rangle \\ & + 2\langle (-i_2(t - \vartheta_{12}))\, u_2(t + \tau_0) \rangle \langle u_1(t) u_2(t + \tau_0) \rangle \end{aligned}$$

$$+ 2\langle(-i_2(t - \vartheta_{12} + \tau_0))\, u_2(t)\rangle\langle u_1(t + \tau_0)u_2(t)\rangle$$

and

$$\begin{aligned}\frac{\partial V}{\partial t_{21}} =& 4\langle -\,(-i_1(t - \vartheta_{21}))\, u_1(t)\rangle\langle u_1(t)u_2(t)\rangle \\ &+ 2\langle(-i_1(t - \vartheta_{21}))\, u_1(t + \tau_0)\rangle\langle u_1(t + \tau_0)u_2(t)\rangle \\ &+ 2\langle(-i_1(t - \vartheta_{21} + \tau_0))\, u_1(t)\rangle\langle u_1(t)u_2(t + \tau_0)\rangle\end{aligned}$$

2.2.3.3 Blind Separation, Nonlinear

Nonlinear Blind Source Separation (NLBSS) refers to a technique used in signal processing to separate multiple signals that have been mixed together, when there is no prior knowledge about the mixing process or the number of sources present. It is considered a challenging problem due to the nonlinearity and complexity of real-world data. We will use an iterative approach, i.e. an algorithm that starts with an initial guess for the source signals and iteratively update them until convergence is reached. They can be computationally expensive, but we are anyway on the street of putting together long formulas.
Motivation
When system equations are nonlinear, a fixed point iteration can be used. Two mixtures of signals (for simplification without time delay) $x_1(t)$ and $x_2(t)$ are received as $y_1(t)$ and $y_2(t)$:

$$\begin{aligned}y_1(t) =& x_1(t) + g^1_{12}x_2(t) + g^1_{212}x_1(t)x_2(t) \\ &+ g^1_{211}x_1(t)x_1(t) + g^1_{222}x_2(t)x_2(t); \\ y_2(t) =& x_2(t) + g^2_{11}x_1(t) + g^2_{212}x_1(t)x_2(t) \\ &+ g^2_{211}x_1(t)x_1(t) + g^2_{222}x_2(t)x_2(t).\end{aligned} \tag{2.21}$$

With the help of the following interaction de-mixing is possible:

$$\begin{aligned}u^n_1(t) =& \frac{1}{1 - h^1_{12}h^2_{11}}(y_1(t) - h^1_{12}y_2(t) \\ &- (h^1_{212} - h^1_{12}h^2_{212})u^{n-1}_1(t)u^{n-1}_2(t) \\ &- (h^1_{222} - h^1_{12}h^2_{222})u^{n-1}_2(t)u^{n-1}_2(t) \\ &- (h^1_{211} - h^1_{12}h^2_{211})u^{n-1}_1(t)u^{n-1}_1(t)); \\ u^n_2(t) =& \frac{1}{1 - h^1_{12}h^2_{11}}(y_2(t) - h^2_{11}y_1(t) \\ &- (h^2_{212} - h^2_{11}h^1_{212})u^{n-1}_1(t)u^{n-1}_2(t) \\ &- (h^2_{222} - h^2_{11}h^1_{222})u^{n-1}_2(t)u^{n-1}_2(t)\end{aligned}$$

$$- (h_{211}^2 - h_{11}^2 h_{211}^1) u_1^{n-1}(t) u_1^{n-1}(t)) \tag{2.22}$$

with a start of the iteration at

$$\begin{aligned} u_1^0 &= 0; \\ u_2^0 &= 0. \end{aligned} \tag{2.23}$$

For $u_1^1(t)$:

$$u_1^1(t) = \frac{y_1(t) - h_{12}^1 y_2(t)}{1 - h_{12}^1 h_{11}^2}. \tag{2.24}$$

For $u_1^2(t)$:

$$\begin{aligned} u_1^2(t) = \Bigg(& y_1(t) - h_{12}^1 y_2 \\ & - \frac{(h_{222}^1 - h_{12}^1 h_{222}^2)(y_2(t) - h_{11}^2 y_1(t))^2}{(1 - h_{12}^1 h_{11}^2)^2} \\ & - \frac{h_{212}^1 - h_{12}^1 h_{212}^2)}{(1 - h_{12}^1 h_{11}^2)^2} \\ & (y_1(t) - h_{12}^1 y_2(t))(y_2(t) - h_{11}^2) y_1(t)) \\ & \frac{(h_{211}^1 - h_{12}^1 h_{211}^2)(y_1(t) - h_{12}^1 y_2(t))^2}{(1 - h_{12}^1 h_{11}^2)^2}) \\ & /(1 - h_{12}^1 h_{11}^2) \Bigg). \end{aligned} \tag{2.25}$$

Of course, Computer Algebra like *Mathematica*-notebook can calculate $u_1^n(t)$ and $u_2^n(t)$:

```
u1[t_Integer]:=(1/(1-h112 h211)) *
(y1-h112 y2
-(h1212-h112 h2212) u1[t-1] u2[t-1]
-(h1222-h112 h2222) u2[t-1] u2[t-1]
-(h1211-h112 h2211) u1[t-1] u1[t-1]
);
u2[t_Integer]:=(1/(1-h112 h211))*
(y2-h211 y1
- (h2212-h211 h1212) u1[t-1] u2[t-1]
- (h2211-h211 h1211) u1[t-1] u1[t-1]
- (h2222-h211 h1222) u2[t-1] u2[t-1]
);
u1[0]:=0;
u2[0]:=0;
```

The h-values have to be adapted such that they are equal to g-values:

With

$$\begin{aligned} h_{12}^1 &= g_{12}^1, \\ h_{11}^2 &= g_{11}^2, \\ h_{212}^1 &= g_{212}^1, \\ h_{212}^2 &= g_{212}^2, \\ h_{222}^1 &= g_{222}^1, \\ h_{222}^2 &= g_{222}^2, \\ h_{211}^1 &= g_{211}^1, \\ h_{211}^2 &= g_{211}^2 \end{aligned} \tag{2.26}$$

we arrive at

$$\begin{aligned} u_1^\infty(t) &= x_1(t) \quad \text{und} \\ u_2^\infty(t) &= x_2(t), \end{aligned} \tag{2.27}$$

if the iteration converges.

Let us assume the g-values are chosen such that there is convergence around the **g**-vector ($\mathbf{g} = (g_{12}^1, g_{11}^2, g_{212}^1, g_{212}^2, g_{222}^1, g_{222}^2, g_{211}^1, g_{211}^2)^T$), for the iteration above with h-values.

Then, for the adaption process of the h values there is a criterium necessary that tells, when the h-values are correct, i.e. equal to g-values. The criterium has to be one-to-one unique: $\mathbf{g} = \mathbf{h}$.

Because the source signals are unknown and only a-priori knowledge is available that the source signals are independent, the only way to construct the criterium is to exploit this independence. If the values of **h** are the same as the values of **g**, then the iteration is having as output $x_1(t)$ and $x_2(t)$, which are the source signals, which are given to be independent. If the values of **h** are not equal to the values of **g**, then u_1^∞ contains terms with $x_2(t)$, e.g. $x_1(t)x_1(t)x_2(t)$, which are not independent of $x_2(t)$. Hence, independence is the right measure for the blind separation task.

Practically, the iteration has to stop somewhere. And independence cannot be checked in all details numerically: Two signals are independent, if all cross-cumulants vanish. With measured signals, independence can only be checked up to a non-infinite order. If e.g. cross-cumulants of order 4 vanish, this does not mean that all cross-cumulants of higher order than 4 vanish. Hence, from vanishing cross-cumulants of order 4 does not follow independence of the signals. If the iteration is stopped, the output contains terms $u_1^{stopIteration}(t)$, in which $x_2(t)$ is still present, but these terms are of a degree of non-linearity that needs to be ignored.

Hence stopping the iteration and deciding to which degree the independence is checked, assumes that neglected terms are of small values.

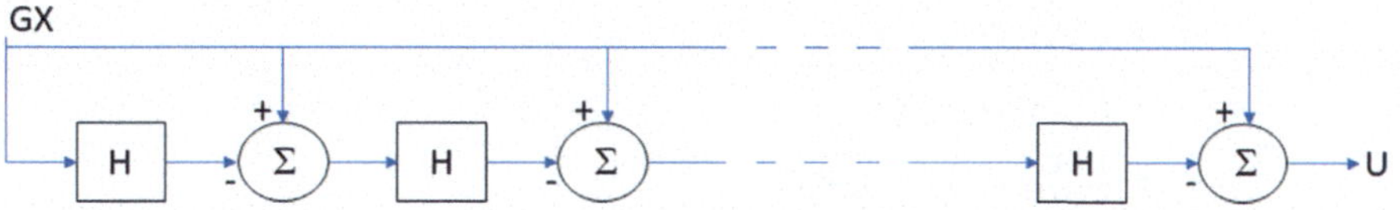

Fig. 2.5 The transformations of the fixpoint equation is realized through sequential execution of Volterra filters

We reformulate the nonlinear blind separation as: If terms of a certain order are neglected, then, when only taking the independence constraint up to the certain order, a set of equations is prepared, such that the number of equations is equal or larger than the number of parameter that need to be calculated.

Furthermore it is important to note that terms like $x_1(t) + x_1(t)x_1(t)$ are independent from $x_2(t) + x_2(t)x_2(t)$. Terms of this kind are the output of the iteration. Hence, the independence criterium is no longer unique. Of course, signal processing methods, or other statistical means, can be used to filter out the nonlinear distortion of the output signals. But the bad news is that, due to the several minima, the cost function is no longer convex, i.e. searching the global maximum becomes a necessity.

General formulation of the Fixpoint Algorithm
The fixpoint iteration can be viewed as a sequential execution of Volterra filters with a finite memory, which can also be written as one polynom (see Fig. 2.5). Hence, we can write Blind Separation as one generic formula, which can then be subject to further operations. In the course of this Book, we have already encountered the BGoF iteration where we stated that we can fuse its iterative steps to one large formula. We will see that the detailness description elements of BGoF (i.e. the order of the Legendre polynom) will be revisited, and that we will use the order of the Legendre polynom as the maximal nonlinearity to be separated by the application of the Blind Separation algorithm in the abstract domain.

Implementation
Whilst the construction of the fixpoint algorithm only contains a few steps, the processing capacity needed to run it on hardware with real data grows very fast with the order of nonlinearity. Special hardware could be used to parallelize processing steps, and the similarity to neural network structures and the polynomial calculations is quite obvious.

However, in the application of Design Space Exploration we will use and exploit this algorithm in an abstract manner. In both cases, the abstract usage and the exploitation in form of signal analysis, we need the algorithm to be available in an analytic formulation, i.e. the output signals have to be presented as a function of the mixture coefficients and the inputsignals, e.g. for the nineth iteration of a cubic separation:

$$\begin{aligned} u_1^9(t) &= k_{11}^{9,1} y_1(t) \\ &+ k_{12}^{9,1} y2(t) \\ &+ k_{211}^{9,1} y_1(t) y_1(t) \end{aligned}$$

$$
\begin{aligned}
&+ k^{9,1}_{212} y_1(t) y_2(t)\\
&+ k^{9,1}_{222} y_2(t) y_2(t)\\
&+ k^{9,1}_{3111} y_1(t) y_1(t) y_1(t)\\
&+ k^{9,1}_{3112} y_1(t) y_1(t) y_2(t)\\
&+ k^{9,1}_{3122} y_1(t) y_2(t) y_2(t)\\
&+ k^{9,1}_{3222} y_2(t) y_2(t) y_2(t)\\
&\vdots \quad \vdots\\
&+ k^{9,1}_{9111111111} y_1(t) y_1(t) y_1(t) y_1(t) y_1(t) y_1(t) y_1(t) y_1(t) y_1(t)\\
&\vdots \quad \vdots\\
&+ k^{9,1}_{9222222222} y_2(t) y_2(t) y_2(t) y_2(t) y_2(t) y_2(t) y_2(t) y_2(t) y_2(t),
\end{aligned}
$$

where multiplications of more than 9 received signals y_i are neglected and where

$$k^{9,1}_{11}, k^{9,1}_{12}, k^{9,1}_{211}, \ldots, k^{9,1}_{9222222222}$$

are functions of the filter

$$h^i_{11}, h^i_{12}, \ldots, h^i_{3222}, (i = 1, 2)$$

of the demixing system: The values for k need to be written as functions of the mixture coefficients.

To check for the independence of the output signals the potential needs to be calculated. The potential is a sum of cumulants of a certain order, whereby the calculation of this order needs the values for the cumulants of a lower order. Hence, the second reformulation is necessary to calculate the statistical values of the output signals $u_i^M(t)$. For example, to calculate $\langle u_1^M(t) u_2^M(t+1)\rangle$ without doing statistic after each change of the demixing coefficients, it is useful to calculate $\langle u_1^M(t) u_2^M(t+1)\rangle$ as function of received signals and mixture coefficients.

Since this results in long chains of algebraic transformations, for example a C++ program can be written,

1. which symbolically performs the multiplications of the iterations and which delivers as result the program to calculate the filter, e.g. for

```
k_9_19111112222 =k_8_111 * k_8_1811112222 * k_9_1211
                +k_8_112 * k_8_1811111222 * k_9_1211
                   ...
                +k_8_211 * k_8_241111 * k_8_242222 * k_9_13222
```

with '*' as a symbolic operator and k_8_... bzw. k_9_... as symbols for the filters;
2. which performs the multiplication of e.g.

$$
\begin{aligned}
&u_1^9(t_1)u_1^9(t_2) \\
&\vdots \\
&u_2^9(t_1)u_2^9(t_2)u_2^9(t_3)u_2^9(t_4)u_2^9(t_5)
\end{aligned}
\tag{2.28}
$$

with symbols, and which sorts the equation such that they have the form 'Filter applied to statistical value'.

In C++ due to the object oriented programming, terms like k_8_... for the time-delayed case are operators.

Furthermore, symmetry feature can be exploited like

$$\langle y_1(t_1)y_1(t_2)y_2(t_3)\rangle = \langle y_1(t_2)y_1(t_1)y_2(t_3)\rangle.$$

After this multiplication step, there is a need to insert brackets around similar terms, such that the formulars can get symbolically shortened.

2.2.3.4 Summary of Blind Separation Technique

The Blind Separation technique is a powerful tool to de-mix received signals, i.e. after Blind Separation there is a possibility to see the values as if they would come directly from the sources (without components of other sources). The Blind Separation technique can work on the basis of polynoms, as an iterative procedure fully automated. There are conditions when Blind Separation with the iterative procedure cannot work, i.e. the Blind Separation iterative procedure needs to be a contraction. In the viewpoint of working with polynoms in the framework of our Design Space Exploration approach, this means that the variables to be separated have to vanish with higher degrees, which is a natural decision making for breaking an iteration. However, state space variables are not to be scalable between zero and one without knowing in advance the scaling parameters. Hence, for the further usage of the Blind Separation technique, we need to identify the right modelling to achieve a generic scaling, which is a necessary (and not yet identified, see Sect. 3.2) tool for the Field Robot: How to build categories? Since we are considering situations for the Field Robot that are only implicitly predictable (they emerge from the system equations, but the programmer does not necessarily see this emergence), we need the Field Robot to have the possibility to generate categories automatically. Only with categories, signals can be analysed by histograms.

Nonlinear separation leads to nonlinearity in the estimation of the source signal, or in other words: There might be several potential solutions for the estimated source signal, which are all inline with the independence criterium. We treat these options as something to be further elaborated on. In nonlinear signal processing, techniques exist for linearizing signals, but here in our later-on abstract application of Blind Separation the linearization might be an assumption that does not fit reality. Hence, we take these options as something that fits into the memory of the processing computer in the Field Robot, and the decision making on which of these options to choose under which conditions is something we are handling later in this Book. We might call the ability of the Field Robot to handle a tree of options as its reasoning component.

In connection of the Bayesian concept, we were discussing the option for an analytical treatment of the equations if we assume Gaussian noise processes. It is possible to take Gaussian noise also in Blind Separation as an input source. It fits to the idea that we break the iterations at a specific level, in which many similar things we do not care about (or we do not know about) are generating a Gaussian noise source. Important for using Blind Separation with Gaussians is to maintain the information flow about the unknown mixing coefficients via the time-delayed correlations (because higher-order statistics does not contain information on Gaussians). Hence, time delayed information needs to be stored inside the Field Robot, which we might call its memory.

Another natural limit of Blind Separation is that a sufficient number of received signals have to be present, or in other words, the number of equations we can write down has to be at least equal to the number of unknowns we need to estimate.

In situations when only a source signal corrupted by noise is received, Blind Separation does not overcome the fact that a decision has to be made: Do we accept the incoming signal or are we assuming that it is just noise? The operation of a Field Robot will be driven by these decision makings, hence, in the next section, we further elaborate on the frequentist strategy, the Bayesian concept, and the Dempster-Shafer theory to handle this decision making.

2.3 Decision Making

In this Section, we apply the Frequentist strategy, the Bayesian concept, and the Dempster-Shafer theory to the problem of decision making. For the frequentist strategy, we echo the Neyman-Pearson Detector, for the Bayesian concept we analyse the Bayesian Channel, and for the Dempster-Shafer theory we re-formulate the example from [3] on inconsistent data fusion. Decision-making is necessary inside the design phase: e.g. Which sensor to put on the robot? Decision-making also happens inside the robot: e.g. Was there a signal received on Sensor A, or was it just noise?

Decision-making is based on the statistics associated with the data and knowledge/uncertainty to use for decision making. Hence, deciding whether a signal is present or not depends on the Signal-to-Noise ratio. Deciding to put a replacement of sensor A on the robot also needs a decision-making base. Certainly, design decisions are made in the light of getting to the best system for performing a given robotic task. Design Space Exploration is preparing the design decision by setting up equations/formulas to optimize, incorporating data, knowledge, uncertainty, and constraints for the design decision making. Decision about the presence of a signal in noise is of a similar abstract mathematical form as design decision-making about which sensor to put on a robot, hence we introduce here the decision-making on signals, already with the outlook to use the algorithms again in a more abstract environment.

2.3.1 Frequentist View: Neyman-Pearson Detector

Following [7], a known signal in White Gaussian noise is detected via the test:

$$\begin{aligned} H_0 &: x[n] = w[n] \qquad n = 0, 1, \ldots, N-1 \\ H_1 &: x[n] = s[n] + w[n] \qquad n = 0, 1, \ldots, N-1 \end{aligned} \tag{2.29}$$

- $s[n]$ assumed to be known with n being the sample number.
  It is important to note, that $s[n]$ here is a digital sample, time is not continuous. For the Field Robot, time will be continuous, hence a design decision about the way the sample is generated has to be made somewhere in the Design Space Exploration. The question on how to handle the differences between continuous and discrete time can also be asked for any other design variable associated to the state space: The grid size to model an object, the histogram bin size to generate statistics for further state machine decision making.
  We further assume
- $w[n]$ is White Gaussian Noise with σ^2.
- $p(x; H_i)$ PDF of x when H_i is true.

where "white" means an autocorrelation function

$$\begin{aligned} r_{ww}[k] =& E(w[n]w[n+k]) \\ =& \sigma^2 \delta[k], \end{aligned} \tag{2.30}$$

in other words: No information is found in time history. This is related to the Field Robot's memory design. If no information in historical time samples can be found, the Field Robot does not need any memory. Definitely, and we will discuss this later on, memory is a critical design variable because the Field Robot needs to extract information about the state of the environment in connection to its own state from a bunch of samples which have to be saved in the Field Robot's memory. Do all sample have to stay there, or only relevant ones, or only recent ones? These are design questions we need to address.

The Neyman-Pearson Detector (NP detector) decides H_1 if the likelihood ratio exceeds a threshold

$$L(x) = \frac{p(x;\, H_1)}{p(x;\, H_0)} > \gamma \tag{2.31}$$

where $x = [x[0]x[1]\ldots x[N-1]]^T$ contains now the N samples of the received signal.

We can calculate the likelihood ratio with

$$\begin{aligned} p(x;\, H_1) &= \frac{1}{(2\pi\sigma^2)^{\frac{1}{2}}} \exp\left[\frac{1}{2\sigma^2}\sum_{n=0}^{N-1}(x[n]-s[n])^2\right] \\ p(x;\, H_0) &= \frac{1}{(2\pi\sigma^2)^{\frac{1}{2}}} \exp\left[\frac{1}{2\sigma^2}\sum_{n=0}^{N-1}x^2[n]\right] \end{aligned} \tag{2.32}$$

and arrive at the decision making

$$L(x) = \exp\left[-\frac{1}{2\sigma^2}\left(\sum_{n=0}^{N-1}(x[n]-s[n])^2) - \sum_{n=0}^{N-1}x^2[n]\right)\right] > \gamma \tag{2.33}$$

or

$$l(x) = \ln L(x) = -\frac{1}{2\sigma^2}\left(\sum_{n=0}^{N-1}(x[n]-s[n])^2 - \sum_{n=0}^{N-1}x^2[n]\right) > \ln\gamma \tag{2.34}$$

Decide H_1 if

$$\frac{1}{\sigma^2}\sum_{n=0}^{N-1}x[n]s[n] - \frac{1}{2\sigma^2}\sum_{n=0}^{N-1}s^2[n] > \ln\gamma \tag{2.35}$$

As $s[n]$ is known, we can calculate

$$T(x) = \sum_{n=0}^{N-1}x[n]s[n] > \sigma^2\ln\gamma + \frac{1}{2}\sum_{n=0}^{N-1}s^2[n] \tag{2.36}$$

$T(x)$ is called "test statistic", i.e., a function of the data

$$T(x) = \sum_{n=0}^{N-1}x[n]s[n] > \gamma' \tag{2.37}$$

We can proceed with a filter design

$$y[n] = \sum_{k=0}^{n} h[n-k]x[k] \tag{2.38}$$

with

$$h[n] = s[N-1-n] \qquad n = 0, 1, \ldots, N-1 \tag{2.39}$$

The resulting algorithm is a correlator: sometimes called “replica correlator”.

In the first discussion here kept simple with $n = 0$ known arrival time.

Hence, the replica correlator or Matched Filter is one example of incorporating knowledge (the signal s) into a signal processing algorithm to fight against noise. Without knowing s we cannot do something against the noise.

2.3.1.1 Frequency Domain

By transforming the calculation into the frequency domain we derive:

$$y[n] = \int_{-\frac{1}{2}}^{+\frac{1}{2}} H(f)X(f)\exp(j2\pi fn)df \tag{2.40}$$

with applying the Fourier transform

$$H(f) = F\{s[N-1-n]\} \tag{2.41}$$

$$H(f) = S^*(f)\exp[-j2\pi f(N-1)] \tag{2.42}$$

we have the same filter $y[n]$ now in the Fourier representation

$$y[n] = \int_{-\frac{1}{2}}^{+\frac{1}{2}} S^*(f)X(f)\exp[j2\pi f(n-(N-1))]df \tag{2.43}$$

or

$$y[N-1] = \int_{-\frac{1}{2}}^{+\frac{1}{2}} S^*(f)X(f)df \tag{2.44}$$

If $X(f) = S(f)$ then $y[N-1] =$ Signal energy, Parseval’s theorem. When we extrapolate the findings for small-band filters to the broadband scenario in real-world Field Robotics, design questions arise on potential weighting of several frequency bands related to their relevance. Same considerations are not only valid time-wise, but also in terms of spatial resolution or in general for the entire state space. Best design choices will always the optimal ones for which we can proof optimality by mathematical technics, one example is given in the next subsection.

2.3.1.2 Optimality of Matched Filter

Again:

$$h[n] = s[N-1-n] \quad n = 0, 1, \ldots, N-1 \tag{2.45}$$

$$h[n] = 0 \quad \text{otherwise} \tag{2.46}$$

Consider all detectors with arbitrary $h[n]$ in $n = 0, 1, \ldots, N-1$ but 0 otherwise.

$$\begin{aligned}\eta =& \frac{E^2(y[N-1]; H_1}{\text{var}(y[N-1]; H_1)} \\ =& \frac{(\sum_{k=0}^{N-1} h[N-1-k]s[k])^2}{E[(\sum_{k=0}^{N-1} h[N-1-k]w[k])^2]}\end{aligned} \tag{2.47}$$

$$\begin{aligned}\eta =& \frac{(h^T s)^2}{E[(h^T w)^2]} \\ =& \frac{(h^T s)^2}{h^T E(ww^T)h} \\ =& \frac{(h^T s)^2)}{h^T \sigma^2 I h} \\ =& \frac{1}{\sigma^2}\frac{(h^T s)^2}{h^T h}\end{aligned} \tag{2.48}$$

Applying the Cauchy-Schwarz inequality

$$(h^T s)^2 \leq (h^T h)(s^T s)) \tag{2.49}$$

we can state that η is maximum if and only if $h = cs$ where c is any constant

$$\eta\text{max} = \frac{s^T s}{\sigma^2} = \frac{\text{Energy}}{\sigma^2} \tag{2.50}$$

If applied to White Gaussian Noise, the Matched Filter maximizes the signal-to-noise ratio (SNR) and is a Neyman-Pearson optimal detector.

For non-Gaussian application the Matched Filter only maximizes the SNR. In the course of the discussions in this Book, we will stick to applying Gaussian noise to signal decision making, which sounds like an unrealistic restriction when considering real-world measurements. Incoming signals on various sensors are by far not Gaussian distributed. But, as we will learn in this Book, we can incorporate modelling knowledge (not only about know signals, but in a more abstract sense about known interactions between the real-world components), such that the understanding of the real-world can be transformed in a Gaussian noise view on signal decision making.

Hence, at the end of the design considerations we have the pay-off of all our extra modelling work, that we can say: And this design is optimal.

2.3.1.3 Performance of Matched Filter

Even more, we can predict the performance of this optimal filter. Aim: P_D for a given P_{FA}. Using replica correlator: decide H_1 if

$$T(x) = \sum_{n=0}^{N-1} x[n]s[n] > \gamma' \tag{2.51}$$

$x[n]$ Gaussian, $T(x)$ is linear combination of Gaussian random variables. $T(x)$ is also Gaussian.

Let $E(T; H_i)$ expected value and $\mathrm{var}(T; H_i)$ variance under H_i.

$$E(T; H_0) = E(\sum_{n=0}^{N-1} w[n]s[n]) = 0 \tag{2.52}$$

$$E(T; H_1) = E(\sum_{n=0}^{N-1}(s[n] + w[n])s[n]) = \text{Energy} \tag{2.53}$$

$$\begin{aligned}
\mathrm{var}(T; H_0) &= \mathrm{var}(\sum_{n=0}^{N-1} w[n]s[n]) \\
&= \sum_{n=0}^{N-1} \mathrm{var}(w[n])s^2[n] \\
&= \sigma^2 \sum_{n=0}^{N-1} s^2[n] = \sigma^2\text{Energy} \qquad (2.54)\\
T' &\sim N(0, 1) \quad \text{under} \quad H_0 \qquad (2.55)
\end{aligned}$$

$$T' \sim N(\sqrt{\frac{E}{\sigma^2}}, 1) \quad \text{under} \quad H_1 \tag{2.56}$$

This leads to the definition of

$$P_{FA} = \Pr\{T > \gamma'; H_0\} = Q(\frac{\gamma'}{\sqrt{\sigma^2 E}}) \tag{2.57}$$

and

$$P_D = \Pr\{T > \gamma'; H_1\} = Q(\frac{\gamma' - E}{\sqrt{\sigma^2 E}}) \tag{2.58}$$

using

$$Q(x) = \int_x^\infty \frac{1}{\sqrt{2\pi}} \exp(-\frac{1}{2}t^2)dt = 1 - \Phi(x) \tag{2.59}$$

$\Phi(x)$ is Cumulative Density Function of $N(0, 1)$.

$\hat{q}^{-1}$ exists because $Q(x)$ monotonically decreasing.

$$\gamma' = \sqrt{\sigma^2 E}\hat{q}^{-1}(P_{FA}) \tag{2.60}$$

$$P_D = Q(\hat{q}-1(P_{FA}) - \sqrt{\frac{\text{Energy}}{\sigma^2}}) \tag{2.61}$$

With more SNR, the argument decreases, and P_D increases.

Increasing the Energy-to-Noise ratio relates to a longer signal and/or a higher amplitude. The shape of the signal does not affect the detection performance.

Plotting P_D against P_{FA} displays the Receiver Operating Characteristic (ROC). The ROC is of high important for the Design Space Exploration approach taken in this Book. Firstly, when designing a Field Robot its performance needs can be given as a point on a mathematical formula similar to a ROC: E.g. the Field Robot shall achieve 90% of its given tasks, declaring only 10% of the scenarios as fulfilled when they are actually not fulfilled.

2.3.1.4 Processing Gain

Giving one example on the effect of the design variable signal length is possible by looking at the input/output relationship for a Matched Filter:

$$\eta_{\text{in}} = \frac{E}{\sigma^2} = \text{for DC level problem } \frac{A^2}{\sigma^2} \tag{2.62}$$

$$\eta_{\text{out}} = \frac{NA^2}{\sigma^2} \tag{2.63}$$

resulting in a Processing Gain

$$\text{PG} = 10\log_{10}\frac{\eta_{\text{out}}}{\eta_{\text{in}}} = 10\log_{10}N\text{dB} \tag{2.64}$$

We see how much Processing Gain we can expect for longer signals. In the Design Space Exploration we have now to understand the balance between time to wait for a Field Robot and the impact of this extra time needed in the light of the operational task given. Furthermore, longer signals have an impact on memory demands, processing power, energy consumptions etc. This simple discussion leads to the insight that a holistic approach for the design of a Field Robot makes sense to be analyzed: Of course, not every intra-system connectivity is of relevance for every task. Some details can be neglected, but which one? This is where a generic approach to Design Space Exploration can help: Can we identify an algorithmic approach to answer these relevance question automatically? Otherwise, i.e. if not automatically, we are left with human experts designing Robots, working in predefined environments only.

2.3.2 Bayes View: Estimation as Decision about a Parameter

Restarting at the previous discussion on discretization of the state space, in this Subsection we use the Bayesian concept to discuss the estimation problem for a continuous variable. The solution to the estimation problem depends on the criterion of goodness by which we measure estimation performance; so, we begin by assigning costs to our decisions about the parameter. In particular, we suppose that there is a function $C : \Lambda \times \Lambda \to R$ such that $C[a, \theta]$ is the cost of estimation a true value of θ as a, for a and θ in Λ. Given such a function C we can then associate with an estimator $\hat{\theta}$ a conditional risk or cost averaged over Y for each $\theta \in \Lambda$; i.e., we have

$$R_\theta(\hat{\theta}) = E_\theta\{C[\hat{\theta}(Y), \theta]\}. \tag{2.65}$$

If we now adopt the interpretation that the actual parameter value θ is the realisation of a random variable Θ, we can define an average of Bayes risk as

$$r(\hat{\theta}) \equiv E\{R_\Theta(\hat{\theta})\}, \tag{2.66}$$

and the appropriate design goal is to find an estimator minimising $r(\hat{\theta})$. Such an estimator is known as a *Bayes estimate* of θ. Noting that $R_\theta(\hat{\theta}) = E\{C[\hat{\theta}(Y), \Theta]|\Theta = \theta\}$, we have

$$r(\hat{\theta}) = E\{C[\hat{\theta}(Y), \Theta]\} = E\{E\{C[\hat{\theta}(Y), \Theta]|Y\}\}. \tag{2.67}$$

By inspection, we see that the Bayes estimate of θ can be found (if it exists) by minimising, for each $y \in \Gamma$, the posterior cost given $Y = y$:

$$E\{C[\hat{\theta}(y), \Theta]|Y = y\}. \tag{2.68}$$

Note that if we assume that Θ has conditional density $w(\theta|y)$ given $Y = y$ for each $y \in \Gamma$, then the Bayes estimate $\hat{\theta}(y)$ corresponding to $y \in \Gamma$ can be sought by minimizing

$$\int_{\Lambda} C[\hat{\theta}, \theta] w(\theta|y) \mu(d\theta). \tag{2.69}$$

We see the integral being used, and we can ask how an representation of a sum of intervals for discrete values of the estimated variable would perform inside the to be explorated Design Space. Can we just say, the variable has two categorical values, like high and low, and the Field Robot handles two different situations only? Or do we need to make a difference of e.g. a height of 1.5 m and 1.45 m? These are design questions that have to be asked and decided upon. Normally, best practice and uncertainty quantification for some questions are used to address these questions. But for a Field Robot in unknown environment there is no best practice and there are many questions due to the complexity of real-world environments. Hence, automatically and holistically approaches again are a good idea (if not even mandatory).

2.3.3 Dempster-Shafer View: Sensor Fusion

A Field Robot might be equipped with some models, producing some prediction or explanation of the processes in the real-world environment. And a Field Robot should have the capability to generate its own modelling knowledge if unknown scenarios are about to be encountered. Hence, a situation has to be handled in which different statements from different models need to be combined. Dempster-Shafer theory can be used for these type of questions, as in the following example.

- t_3 no replica signal present
- t_4 no replica signal present but instead a very large signal later
- t_1 replica signal present
- t_2 a similar signal present

In case of t_3 and t_4 no signal is present, otherwise a signal is present.

Suppose we are asking three sensors.

We are receiving the following reports:

- First sensor "I am 60% certain that either t_3 or t_1 is true."
- Second sensor "I am 70% sure that either t_3 or t_4 is true."
- Third sensor "I am 80% sure that one of the four statements is right, where
  - $\{t_2\}$ is twice as likely as $\{t_1\}$
  - $\{t_3\}$ is three times as likely as $\{t_1\}$
  - $\{t_4\}$ is twice as likely as $\{t_1\}$

Now, Dempster-Shafer approach is used to combine these statements.

- First sensor: see Table 2.2
- Second sensor:

$$m_2(S) = 0.3$$

$$m_2(\{t_3, t_4\}) = 0.7,$$

see Table 2.3.
The important feature of the Dempster-Shafer Theory is that a Sum of Inconsistencies k gives a normalisation factor

$$0.634 = 1 - \sum k \tag{2.70}$$

Result for $m(\{t_3\})$

$$m(\{t_3\}) = \frac{0.084 + 0.084 + 0.126 + 0.054 + 0.036}{0.634} = 0.606 \tag{2.71}$$

The result can be expressed as "Support-Plausibility-Intervals", giving a Field Robot values to optimize for in its operation, and for the Designer we have an analytical formula to insert in the calculations modelling the Field Robot (Tables 2.2 and 2.3).

Table 2.2 Data fusion in the Dempster-Shafer view: first sensor

Event	Support	Plausibility
$\{t_1, t_3\}$	0.6	1.0
S	0.4	1.0

Table 2.3 Data fusion in the Dempster-Shafer view: second sensor

	$m_2(\{t_3, t_4\}) = 0.7$	$m_2(S) = 0.3$
$m_1(\{t_1, t_3\} = 0.6$	$m(\{t_3\}) = 0.42$	$m(\{t_1, t_3\}) = 0.18$
$m_1(S) = 0.4$	$m(\{t_3, t_4\}) = 0.28$	$m(S) = 0.12$

2.3.3.1 Summary of Decision Making

The important take-away is that the Matched Filter is an optimal filter for Gaussian noise, that we can incorporate cost values into Bayesian decision making (whereby the cost/risk values relate to predefined or predicted values), and that the Dempster-Shafer theory allows us to deal with inconsistencies.

For the Field Robot these take-aways might have the following impacts: If the robot comes to an (abstract) situation where all it can do is to ignore the noise and take the received signal as the sufficient information projected on the cost regarding the given task, all further reasoning would be without effect. Cost prediction and cost-value predefinition relates to game theory, also in an abstract view (not only in a game with an opponent), and we will use this relationship further when we develop an automated version of Design Space Exploration. The inconsistency values are a natural communication item between an autonomous robot and a human: Signalling that something is not as it should be, gives the human insight in the operation without getting overwhelmed with all possibly unstructured information. Also inside the robot, the inconsistency plays a crucial role when designing update cycles for the robot's own models. This will be discussed in Chap. 5.

Take-Away for Design Space Exploration

As Fig. 2.4 depicts, the three relationships to the sample in the interface can be combined by the methods BGoF, FISST, and BSS. By the combination we are able to reach all abstraction levels, both semantically and syntactically, and relate the sample value to them.

A central role plays the Bayesian concept, which allows to memorize (to write down) syntactically the findings on interactions (categorically in terms of classification and time-wise in terms of historical or future tracks) with the help of the inherent metric of probability (i.e. sum over all options equals one). Especially the view on always accepting a latent layer below the currently memorized layer is related to the autonomy approach for a field robot: Is the latent layer to be explored next, or is the current memory entry sufficient to fulfil the given task?

References

1. Schuster, H.G., Ehlers, F.: Signaltrennungsverfahren und -einrichtung für nichtlineare Mischungen unbekannter Signale, DE19531388C1 (1995)
2. Mahler, R.: Random set theory for target tracking and identification. In: Handbook of Multisensor Data Fusion. CRC Press (2001)

3. Carl, J.: Contrasting approaches to combine evidence. In: Handbook of Multisensor Data Fusion. CRC Press (2001)
4. Ehlers, F.: Particle filter for information management in an autonomously operating activated distributed sensor network. In: Proceedings of OCEANS Conference 2007, Aberdeen (2007)
5. Mukhopadhyay, S., Fletcher, D.: 2018 paper Generalized Empirical Bayes via Frequentist Goodness of Fit (2023). https://www.nature.com/articles/s41598-018-28130-5. Accessed from 15 Jan 2023
6. Koch, W.: Tracking and Sensor Data Fusion. Springer (2014)
7. Poor, H.V.: An Introduction to Signal Detection and Estimation. Springer (1994)
8. Ehlers, F.: Quellenseparation [Nonlinear Blind Source Separation], Thesis, Dec. 1997
9. Gelman, A., Carlin, B., Stern, H., Dunsun, D., Vehtari, A., Rubin, D.: Bayesian Data Analysis. CRC Press, 3rd edn (2014)
10. Ehlers, F., Schuster, H.G.: Blind separation of convolutive mixtures and an application in automatic speech recognition in noisy environment. IEEE Trans. Signal Process. **45**(10), 2608 (1997)

Chapter 3
Physics

Abstract Real robots are called embodied. And this embodyment makes a big difference to simulated robots. A simulated robot is still a program, living so to say inside the computer, also if hardware is in the loop that is calculating based on simulated input data. A real robot is receiving real signals with no ground truth readable in some memory cells of a computer. Information about a real state is only coming from better sensor, and quantum mechanics is telling us that also a better sensor has fundamental limits. Whilst in a computer program one can work with semantically true for logical program outputs, the semantics (i.e. meaning) in the embodied world needs an underlying truth to which the robot has no access to. Instead the robot has to build a model, the model is implemented as computer program, and there: The robot can decide on its own for a semantic truth. This decision has no direct effect on the semantic truth in the real world. How does such model inside the robot look like? In its most general form it is a set of equations describing a random dynamical system. Also the Designer has to work with models for the semantic truths. It seems to be logically necessary that the models in the (to be designed) robot and in the Design Space Exploration phase are assumed to be the same. Whilst the robot works with data, in the design phase we do neither have a robot nor data. Hence, we need a method to produce as much data as necessary inserting our model knowledge into an adequate sampling method. In this Book, we choose the Polynomial Chaos Expansion (PCE) as a data generation method. From an analytical point of view, we transform the equations describing the random dynamical system into a sequence of samples, but instead of performing the sampling, we only further investigate the syntactical shape of the transformation equations, and we are extracting from these equations the moments of the pdfs these equations are describing. We find, that we can successively formulate the random dynamical system as a cascade of increasing detailedness with each layer in this hierarchical system of decision-making driven by Gaussian noise. We further link to analytical tools to describe and further elaborate on the Gaussian noise driven decision-making scenario: Stochastic Optimal Control, Koopman Operator, Hidden Markov Model and its analytical pendant (the Integrated Nested Laplace Approximation).

F. Ehlers, *Design Space Exploration in Robotics*, Studies in Systems, Decision and Control 229, https://doi.org/10.1007/978-3-031-78740-9_3

Top-Down View

We can use the results of the Signals chapter to build/estimate a model as a Random Dynamical System (RDS). With the help of the Polynomial Chaos Expansion (PCE) technique we can simulate samples from this model, as is depicted in Fig. 3.1.

However, to do this efficiently we need to further work on this simulation: Randomness comes in with stochastic influence every time step and with uncertainty about model parameters, the first Gauss-distributed (due to many lower level interactions), the latter as uniformly distributed (we just know the interval in which values can be found). But PCE is only working with one type of noise processes to model, hence we need to decide whether to model with Legendre or Hermite polynomial, see Fig. 3.2.

We can use Blind Source Separation (BSS) now in the conceptual realm, and separate into a Gaussian source and a Uniform source of uncertainty, as depicted in Fig. 3.3.

We identify the Design Space as a tree of options, in which H^∞-control, the Koopman operator, Hidden Markov Models (HMM), and the INLA algorithm, as further mathematical tools, describe how the Field Robot moves/behaves within the Random Dynamical System.

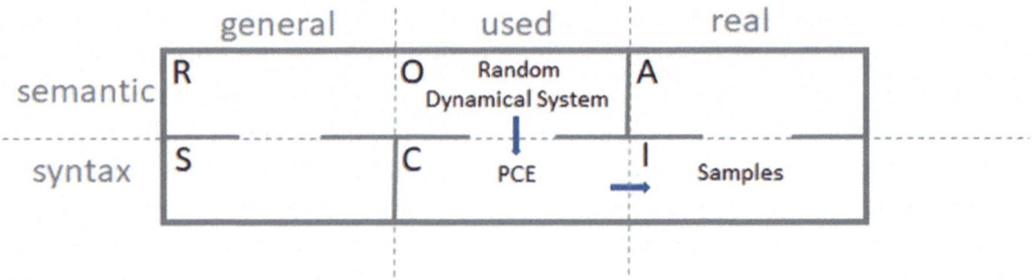

Fig. 3.1 The Polynomial Chaos Expansion is capable to generate samples from a description of a Random Dynamical System

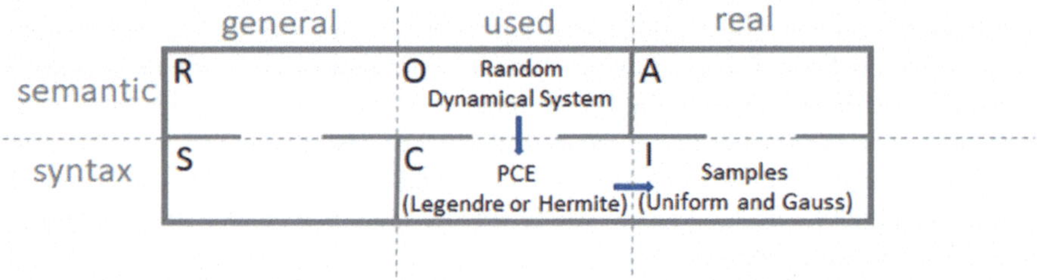

Fig. 3.2 Samples at the interface are generated by both, Gaussian and Uniform randomness. But the PCE expansion only works for either Uniform modelling with the Legendre polynom or Gaussian modelling with the Hermite polynom

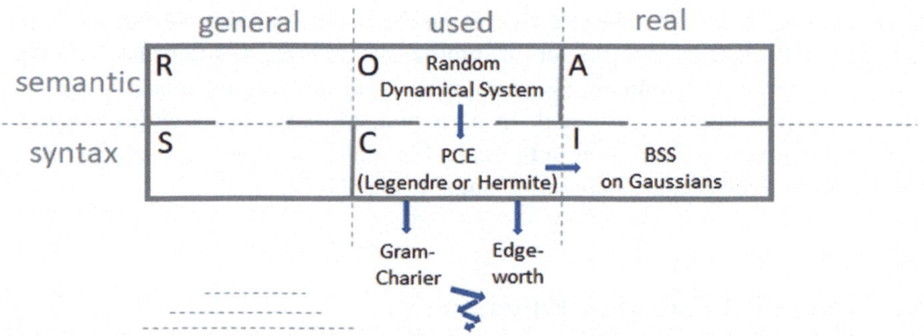

Fig. 3.3 If we perform a layered separation into a Gaussian source and a Uniform source of uncertainty, we can restrict the signal processing and control to Gaussian treatment

Following story from the first part of the Introduction, when we continue to analyse signals from the outer world we can understand their sources, i.e. we can model what is going on in this (from the to be designed robot's point of view) outer world. However, we will always have a limited precision on what is going on there, and even more, just like we receive only noisy signals, the outer world is driven by noisy signals. The outer world turns out to be quite large and describing all the sources, their interconnections, inter-dependencies, etc. needs some organisation (things to service match), also if we want to share our information with others. But this organisation is not fixed, it depends on our application task: our operation goal which is the reason why the robot is listening to signals, why it is collecting information. The robot needs to reach the given operational goal and it needs to be careful over time not to run into difficulties about its ability to reach it, i.e. it has to control the situation.

3.1 Equations to Describe Random Dynamical Systems

A random dynamical system (RDS) is a mathematical framework that describes the evolution of stochastic processes through time, governed by random differential equations or difference equations. The approach unifies concepts from deterministic and probabilistic dynamics, providing a versatile tool for modeling complex systems in various fields such as physics, biology, finance, and engineering. RDS allows researchers to study the long-term behavior of stochastic processes, analyze their stability properties, and develop efficient algorithms for computational purposes, thus contributing significantly to our understanding of dynamic systems under uncertainty or randomness.

From the many ways to write down these equations, we suggest two for the interested reader: Random Ordinary Differential Equations and Random Markov Chains [16]. Random Ordinary Differential Equations (RODEs) often generate random dynamical systems. These and their attractors, called random attractors, are living in state space R^d. Markov chains can be used to model the random environments and other random factors in biological systems, as often the switching between different environments is memoryless and the waiting time for the next switch is exponentially distributed. In particular, Markov chains with tridiagonal transition matrices are common in biological models, for example, birth-and-death processes, cell-cell communication, to name just a few.

3.2 Cascaded Gaussian Functionals

Remember, we prepare the development of a generic metric for the robotic autonomy. We see the robot as a system that can be described by a random differential equation. Parameters are unknown, physics can be modelled. If we solve the equation, we can add to the solution an optimization process e.g. for robustness and efficiency with constraint to reach the given target. Certainly, solving the world's equation as a whole is infeasible, but if we are able to iterate through different levels of importance or relevance, we might be able to link the metric for robotic autonomy with observations regarding the way we step through these levels of importance or relevance.

As we have seen in the previous section, discrete and continuous processes are both tools to analyse Random Dynamical Systems. Hence, the idea is to look for cascading the levels of importance of detailness through the entire Random Dynamical System, switching between continuous levels which we assume to be driven by Gaussian and uniform noise and levels in which discrete decision-making based on categories and attracted discrete status values is happening. In other words, the idea is to look from a decision-making (outside) level on a continuous level (as inside), and then stepping through the inside if we decide that something of relevance (for the given task) is likely to be found in the details of this inside continuous level.

3.2.1 Separation of Equations

The question is, how can we find out whether something interesting is happening in the inside continuous level? We need to solve the system's equation first to find out. But the system's equation include randomness on the parameters as well as randomness in every time step due to Gaussian noise. The first step would then to transform the time-step-wise Gaussian noise by suitable mathematical approaches into a before-or-after description of unknown Gaussian random effects, just as we include the randomness of unknown parameters. Then, we have two different randomness processes adding to the equations: uniform randomness and Gaussian randomness. Then, we might be able to separate these two sources of randomness, as we have treated nonlinearly mixed signals in the chapter before. In fact, these ideas fit very well to a mathematical technique, called Polynomial Chaos Expansion, which we introduce next.

3.2.1.1 Polynomial Chaos Expansion (PCE)

Random Differential Equation
In [6], a Galerkin-based Solution of Random Differential Equation is given. The tutorial demonstrates how random differential equations can be solved using polynomial chaos expansions (PCE). The tutorial explains the theory with an example. A random differential equation is an ordinary differential equation that has random parameters, hence its solution is itself a (time-varying) random variable. Perhaps the simplest non-trivial example is the following scalar, linear ordinary differential equation

$$\dot{x}(t) = ax(t), x(0) = x_0, \tag{3.1}$$

where a is the realization of a Gaussian random variable $a \sim N(\mu, \sigma^2)$ with mean μ and variance σ^2.

The form of the differential equation can be different, but the treatment with PCE is generally the same. We have already seen in the previous chapter, that long polynomial formulas need to be treated when performing the Design Space Exploration with the help of analytical treatment. On the other hand, analytical treatment is the only solution if no data is present, and in the design phase there is no system producing data. Hence, we work with the only chance we have, i.e. analytical treatment, although is will be symbolically computational heavy. Later, we will see that we can organize our formulas well and that we can step forward with a transformation of long detailed equations to a macroscopic description of the behavior or capability of the Field Robot.

Arguably, for every realization a we can solve the differential equation and obtain $x(t) = x_0 e^{at}$, from which we find that

$$ln(x(t)) = ln(x_0) + at \sim N(ln(x_0) + \mu t, (\sigma t)^2). \tag{3.2}$$

In other words, the logarithm of the solution is normally distributed (so-called log-normal distribution). We'd like to obtain this result numerically with the help of PCE. The first step is to define the (truncated) PCE for the random variable a

$$a = \sum_{i=0}^{L} a_i \phi_i, \tag{3.3}$$

where a_i are the so-called PCE coefficients, and ϕ_i are the orthogonal basis polynomials. As the solution to the random differential equation is itself a random variable, we treat $x(t)$ as the realization of the random variable $x(t)$, and define its PCE

$$x(t) = \sum_{i=0}^{L} x_i(t) \phi_i. \tag{3.4}$$

The question is how to obtain the unknown PCE coefficients $x_i(t)$ from the known PCE coefficients a_i relative to the orthogonal basis polynomials ϕ_i. This can be done using Galerkin projection, which is nothing else than projecting onto the orthogonal basis. Think of a three-dimensional space, in which you have placed some three-dimensional object. If you now project the silhouett of the object onto every axis of the three-dimensional space, then you are doing a Galerkin projection. With PCE the concept is equivalent, but the imagination has a harder time. The first step for Galerkin projection is to insert the PCEs the second step is to project onto every basis polynomial ϕ_m for $m = 0, 1, \ldots, L$, and to exploit orthogonality of the basis. Of course, the initial condition must not be forgotten: $x_0(0) = x_0, x_m(0) = 0$ with $m = 1, \ldots, L$. If we can solve such enlarged system of ordinary random differential equations, we can reconstruct the analytic solution.

Stochastic Partial Differential Equations
Let's give a more complex example how to use PCE when dealing with a model of the world given by systems of differential equations. Following [3], the derivation of the system of stochastic PDEs governing linear acoustic wave propagation begins with the deterministic system. The deterministic system of equations we choose to use is the coupled first-order system of partial differential equations called the velocity-pressure system

$$\frac{\partial v_x(x,t)}{\partial t} = -\frac{1}{\rho(x)}\frac{\partial p(x,t)}{\partial x} + \frac{1}{\rho(x)} s_x(x,t) \tag{3.5}$$

$$\frac{\partial v_y(x,t)}{\partial t} = -\frac{1}{\rho(x)}\frac{\partial p(x,t)}{\partial y} + \frac{1}{\rho(x)} s_y(x,t) \tag{3.6}$$

$$\frac{\partial v_z(x,t)}{\partial t} = -\frac{1}{\rho(x)}\frac{\partial p(x,t)}{\partial z} + \frac{1}{\rho(x)} s_z(x,t) \tag{3.7}$$

$$\frac{\partial p(x,t)}{\partial t} = -\kappa[\frac{\partial v_x(x,t)}{\partial t} + \frac{\partial v_y(x,t)}{\partial t} + \frac{\partial v_z(x,t)}{\partial t}] + \frac{\partial m(x,t)}{\partial t} \tag{3.8}$$

where v_x, v_y, and v_z are the particle velocity components; p is pressure; ρ and κ are the medium density and bulk modulus, respectively. All of the dependent variables (v_x, v_y, v_z, and) are functions of 3-D space (x) and time (t), whereas we are assuming here that the medium parameters (ρ and κ) are functions of 3-D space only. Sources (s_x, s_y, s_z, and m) are also functions of space and time and are defined by

$$s_x(x,t) = f_x(x,t) + \frac{\partial m^a_{xx}(x,t)}{\partial x} + \frac{\partial m^a_{xy}(x,t)}{\partial y} + \frac{\partial m^a_{xz}(x,t)}{\partial z} \tag{3.9}$$

$$s_y(x,t) = f_y(x,t) + \frac{\partial m^a_{xy}(x,t)}{\partial x} + \frac{\partial m^a_{yy}(x,t)}{\partial y} + \frac{\partial m^a_{yz}(x,t)}{\partial z} \tag{3.10}$$

$$s_z(x,t) = f_z(x,t) + \frac{\partial m^a_{xz}(x,t)}{\partial x} + \frac{\partial m^a_{yz}(x,t)}{\partial y} + \frac{\partial m^a_{zz}(x,t)}{\partial z} \tag{3.11}$$

$$m(x,t) = -\frac{1}{3}[m^s_{xx}(x,t) + m^s_{yy}(x,t) + m^s_{zz}(x,t)] \tag{3.12}$$

where m_{ija} are components of the antisymmetric portion of the 3 by 3 moment tensor; miis are the diagonal components of the symmetric portion of the moment tensor; f_x, f_y, and f_z are components of a body force. m then is simply an isotropic pressure source like an explosion or implosion. Typically, all m_{ija} are zero, so that only the body force components will act as sources for the velocity equations.

The first stage in deriving the random acoustic PDEs is to expand all dependent and independent variables as sums over orthogonal basis functions of random variables, which are assumed to be independent of both space and time. For reasons that will be explained below, the orthogonal basis functions used are taken from a class of orthogonal polynomials called chaos polynomials. The dependent and independent variables become

$$v_i(x,t) = \sum_{j=0}^{N} V_{i,j}\Phi_j(\theta) \tag{3.13}$$

$$p(x,t) = \sum_{j=0}^{N} P_j\Phi_j(\theta) \tag{3.14}$$

$$\frac{1}{\rho(x,t)} = \sum_{j=0}^{N_r} R_j(x,t)\Phi_j(\theta) \tag{3.15}$$

$$\kappa(x,t) = \sum_{j=0}^{N_\kappa} K_j(x,t)\Phi_j(\theta) \tag{3.16}$$

$$s_i(x,t) = \sum_{j=0}^{N_{s_i}} S_{i,j}(x,t)\Phi_j(\theta) \tag{3.17}$$

$$m(x,t) = \sum_{j=0}^{N_m} M_j(x,t)\Phi_j(\theta) \tag{3.18}$$

In these equations, i is an index variable and stands in place of (x, y, or z); V_i, j, S_i, j, P_j, R_j, K_j, and M_j are the expansion coefficients for the chaos polynomial basis function of order j, $\Phi_j(\theta)$, of random variable(s). Note that the buoyancy (reciprocal of density) is expanded instead of density itself. This simplifies the derivation of the equations below. Note that there are five different maximum orders in the sums for the Equations: N, N_r, N_k, N_{s_i}, and N_m. The latter four are needed for the independent variable expansions. They are independent of each other and are dependent solely on the maximum chaos polynomial order required to adequately represent the underlying probability distribution for that medium parameter or source. As will be discussed in more detail, one chooses which particular class of chaos polynomials to use based on the one that will represent the medium parameter and/or source probability distribution with the fewest number of expansion coefficients. If any of these maximum orders are zero, then it is equivalent to assuming that parameter is known exactly. The maximum order for the dependent variables, N, is user adjustable. To be formally correct, except in certain very simple situations, would need to be infinite; however, we choose to truncate the expansion at some N. How large N needs to be depends on the probability density functions, the class of chaos polynomials chosen, the number of independent random variables, and the set of PDEs being solved. The cascaded approach in our design space exploration approach tells us the best choice for the value of N. Hence, we can assume some N has been suitably chosen. To proceed with the derivation, we place the PCE into the differential equations, recalling that the chaos polynomial basis functions are independent of time and space

$$\sum_{j=0}^{N} \frac{\partial V_{x,j}}{\partial t} \Phi_j(\theta) = -\sum_{i=0}^{N}\sum_{j=0}^{N_r} R_j \frac{\partial P_i}{\partial x} \Phi_i(\theta)\Phi_j(\theta) + \sum_{i=0}^{N_{sx}}\sum_{j=0}^{N_r} R_j S_{x,i} \Phi_i(\theta)\Phi_j(\theta) \tag{3.19}$$

$$\sum_{j=0}^{N} \frac{\partial V_{y,j}}{\partial t} \Phi_j(\theta) = -\sum_{i=0}^{N}\sum_{j=0}^{N_r} R_j \frac{\partial P_i}{\partial y} \Phi_i(\theta)\Phi_j(\theta) + \sum_{i=0}^{N_{sy}}\sum_{j=0}^{N_r} R_j S_{y,i} \Phi_i(\theta)\Phi_j(\theta) \tag{3.20}$$

$$\sum_{j=0}^{N} \frac{\partial V_{z,j}}{\partial t} \Phi_j(\theta) = -\sum_{i=0}^{N}\sum_{j=0}^{N_r} R_j \frac{\partial P_i}{\partial z} \Phi_i(\theta)\Phi_j(\theta) + \sum_{i=0}^{N_{sz}}\sum_{j=0}^{N_r} R_j S_{z,i} \Phi_i(\theta)\Phi_j(\theta) \tag{3.21}$$

$$\sum_{j=0}^{N} \frac{\partial P_j}{\partial t} \Phi_j(\theta) = -\sum_{i=0}^{N}\sum_{j=0}^{N_\kappa} K_j \left[\frac{\partial V_{x,i}}{\partial x} + \frac{\partial V_{y,i}}{\partial y} + \frac{\partial V_{z,i}}{\partial z}\right] \Phi_i(\theta)\Phi_j(\theta) + \sum_{i=0}^{N_m} \frac{\partial M_i}{\partial t} \Phi_i(\theta) \tag{3.22}$$

Now we determine the inner product of each equation with $\Phi k(\theta)$ using the definition of inner product $f(x)g(x) = \int_a^b f(x)g(x)w(x)dx$ where $w(x)$ is the weight function specific to the chosen chaos polynomial and a and b confine the integration to the support space of the polynomial. Applying this with $\Phi_k(\theta)$ and assuming that we have normalized the basis functions relative to this inner product definition so that we have orthonormal chaos polynomials, we can use the orthonormality of the basis functions to obtain

$$\frac{V_{x,k}}{\partial t} = -\sum_{i=1}^{N}\sum_{i=1}^{N_r} e_{ijk} R_j \frac{\partial P_i}{\partial x} + \sum_{i=1}^{N_{sx}}\sum_{i=1}^{N_r} e_{ijk} R_j S_{x,i} \tag{3.23}$$

$$\frac{V_{y,k}}{\partial t} = -\sum_{i=1}^{N}\sum_{i=1}^{N_r} e_{ijk} R_j \frac{\partial P_i}{\partial y} + \sum_{i=1}^{N_{sy}}\sum_{i=1}^{N_r} e_{ijk} R_j S_{y,i} \tag{3.24}$$

$$\frac{V_{z,k}}{\partial t} = -\sum_{i=1}^{N}\sum_{i=1}^{N_r} e_{ijk} R_j \frac{\partial P_i}{\partial z} + \sum_{i=1}^{N_{sz}}\sum_{i=1}^{N_r} e_{ijk} R_j S_{z,i} \tag{3.25}$$

$$\frac{\partial P_k}{\partial t} = -\sum_{i=0}^{N}\sum_{i=0}^{N_\kappa} e_{ijk} K_j \left[\frac{\partial V_{x,i}}{\partial x} + \frac{\partial V_{y,i}}{\partial y} + \frac{\partial V_{z,i}}{\partial z}\right] + \frac{\partial M_k}{\partial t} \tag{3.26}$$

where $e_{ijk} = \Phi_i \Phi_j \Phi_k$ (which is known and can be tabulated for each i, j, k triple beforehand). This system of equations defined form the stochastic PDEs for an acoustic medium. These are similar in form to the deterministic wave equations; indeed, if one assumes that all independent variables are exactly known (equating $r = k = N_{s_i} = N_m = 0$) then one recovers the deterministic equations as one should. Despite their similarity, it is obvious that there are many additional terms in each equation and that there are sets of the four equations that must be solved, one set for each index k. However, the number of additional terms is not as great as one might first think. e_{ijk} tends to be relatively sparse so many of the additional terms in the sums are trivially zero.

Summary on PCE
The important observation is that z in the PCE is selected between zero and one, hence suitable for a Blind Separation technique. What, if we take the Random Dynamical System and write it as a mixture of two PCE-Processes: A Legendre polynomial for uniform probability density function and an Hermite polynomial for Gaussian. Let us elaborate this idea in the following discussion.

3.2.1.2 Requirement View versus Specification View

When the robot gets its order to perform a specific task, the goal is described with an indicator for a sufficient result of its operation. Knowing that there is no exact 100% solution, the order contains as requirement a working point on the receivers operating characteristic (ROC), which we have first encountered in the Decision Making section. Hence, we can associate the requirement formulation to a modelling with Hermite polynoms. Furthermore, if we specify a scenario in which the robot has to deal with a-priori unknown settings, honestly (or naturally) we would model this by an almost complete ignorance, i.e. a uniform interval, represented in PCE by Legendre polynoms.

We can now write down equations to describe the states of the robot and the environment for state space variables x and impulse p, following the standard procedure in Hamiltonian mechanics. Then, we have two different cases:

- a modelling where we can estimate x through an ignorance of Gaussian noise influence, i.e. a Matched Filter like situation, and where we can control x in the presence of Gaussian noise with a certain amount of invested energy: In this situation, analytic solutions for the best filter and the best controller are available, and as the Gaussian noise influence is assumed to be vanished (in projection of the given task has become irrelevant), we can use PCE's Legendre polynoms to generate samples for the system with solved system equations.
- a modelling where the Gaussian noise is driving the system and the Design Space Exploration can set a control mechanism for the system's behavior, just making sure that the system will get into a state described by some interval limits. State space and impulses in this case are mixtures of PCE processes related to Gaussian and uniform probability density functions, whereby the Legendre polynoms are

available from the previously discussed case, i.e. solutions of the associated equations of a (sufficiently controlled) estimation process with sufficient performance for the task given to the robot.

The important observation from the discussion above is that a hierarchical process connects the two cases. Let us inspect this from the bottom-up view: Assume the equations for Decision Making and Control (we look deeper into the control equations in the subsection Analytical Tools) are solved with a sufficient ROC performance. Then, we can insert this solution into the Gaussian driving noise representation, where we have two equations, one for the state space and one for the impulses.

3.2.1.3 Separation of Canonical Equations for Requirements

In both equations, there is a nonlinear mixture of the PCE samples for uniform and Gaussian processes: This is a Blind Separation situation. Samples from two different independent sources, also scaled between zero and one, hence the iterative nonlinear fixpoint solution is applicable. Although the iterative nonlinear fixpoint solution is just a technique here without any source signal separation aspects, there is an interpretation of this abstract approach: If the Design Space Exploration is seen as a non-cooperative game between requirements and specifications, the nonlinear separation calculates the best solution for fulfilling the requirements with the widest range of specifications.

After separation of Hermitian and Legendre descriptions, we can use further analytical tools to analyse the solutions, which is also necessary, because a Blind Separation has a built-in permutation: Which output belongs to which process has to be identified in a post-processing step, which we will have a look at when we discuss parametrisations of pdfs.

As we are in the framework of using polynomials to describe all functional relationships, we can as a generic approach consider all solutions of differential equations to be found by comparisons of coefficients of Taylor expansions. Of course, there are more efficient calculatoric approaches, e.g. as in the PCE section for the sound propagation where the wave equation is already known to be a solution. But the Taylor expansion approach fits well into the overall approach of analyzing equations only up to a certain degree of resolution.

3.2.1.4 Hierarchical Version

When we continue the bottom-up perspective, then we can see that it is possible to iteratively build on the finding for the lowest level, quasi transporting the information and controlled situation into the next hierarchical level where we apply the same mechanism of Blind Separation. Hence, once we have found a lowest level in the hierarchy, we can build-up a complete hierarchy in which each layer prepares the situation and communicates information to the higher layer.

We need the top-down perspective to identify the lowest levels for different situations the Field Robot might fit itself in. Hence, we inspect the hierarchical version from the top requirement level: We play the game requirement against specifications with a certain resolution (i.e. the highest degree of the Chaos Polynomials used), solve the equations, find the options for system design from the nonlinear solution, and for each option we need to decide whether this solution is sufficient for the given task. If it is not sufficient, we have the option to increase the resolution: Take a higher degree of the Chaos Polynomials used. We can then re-use the previous separation results, because we only have to add the separation for the higher polynomial degrees in the associated fixpoint equations. For each option we continue this process until we reach a situation where from this lowest level the resulting ROC is sufficient for the given task, whereby the resulting ROC is calculated from the bottom-up calculations just discussed. This top-down/bottom-up process assures that the two directions lead to an identical mathematical structure of the equations: The effort of looking for a complete analytical solution pays off now, where we have (very long) polynomial equations for different options.

There is more algorithmic effort behind the scene in the sense that global optimization results have to be found. Since we are purely analytic in the polynomial regime, calculating derivates and checking for local optima, is assumed to be viable in a computer-aided algebra program.

Next, we will have a short look at how to solve the permutation problem in Blind Separation for our abstract application of the technique. Then, we summarize the finding in the section Design Space as a Tree of Options, because on this tree and its relationship to the analytic equations just derived we will focus the discussion when discussing the Design Space Exploration in Robotics.

The remainder of the chapter is then about mathematical techniques we mentioned in the previous discussion. (In order to keep the previous discussion concise they have been left out there, but it is important to take a look at them because we will need to model the capacity of the Field Robot, hence we need to see how it can perform its internal calculations which have to be linked to the Design Space Tree of Options.)

3.2.1.5 Parametrisation of Pdfs

When we are working with the PCE processes for Legendre and Hermite polynoms, we want to arrive at a uniform distribution for the Legendre polynoms and a Gauss distribution for the Hermite polynoms. It is possible to analytically derive moments of the distributions from PCE representations without sampling [15]. As we will echo with the following few sentences, Gram-Charlier representations cannot be used to render Gaussian distributions. Hence, we have an algorithmic approach to solve the Blind Separation permutation-issue, i.e. matching the best representations to the outputs of the Blind Separation.

Let us look into Gram-Charlier and Edgeworth expansions [7] in more detail.

The Gram-Charlier A series and the Edgeworth series are series that approximate a probability distribution in terms of its cumulants. The key idea of these expansions is to write the characteristic function of the distribution whose probability density function f is to be approximated in terms of the characteristic function of a distribution with known and suitable properties, and to recover f through the inverse Fourier transform.

For the approach to Design Space Exploration in this Book, the (rather detailed) discussion of the derivation of the Gram-Charlier A series and of the Edgeworth series, gives also a further insight into the way the analytical calculations to be performed will look like: treatment of various polynomials with an internal periodicity and cumulants generation, plus the expansion with regard to the normal distribution (for which we have already stated in the Signals chapter that we need to exploit the availability of proofs for optimality when aiming for a maximized trustworthiness of the to be designed Field Robots).

Gram-Charlier A series
We examine a continuous random variable. Let $\hat{f}$ be the characteristic function of its distribution whose density function is f, and κ_r its cumulants. We expand in terms of a known distribution with probability density function ψ, characteristic function $\hat{\psi}$, and cumulants γ_r. The density ψ is generally chosen to be that of the normal distribution, but other choices are possible as well. By the definition of the cumulants, we have

$$\hat{f}(t) = \exp\left[\sum_{r=1}^{\infty} \kappa_r \frac{(it)^r}{r!}\right] \text{ and} \tag{3.27}$$

$$\hat{\psi}(t) = \exp\left[\sum_{r=1}^{\infty} \gamma_r \frac{(it)^r}{r!}\right], \tag{3.28}$$

which gives the following formal identity:

$$\hat{f}(t) = \exp\left[\sum_{r=1}^{\infty} (\kappa_r - \gamma_r) \frac{(it)^r}{r!}\right] \hat{\psi}(t). \tag{3.29}$$

By the properties of the Fourier transform, $(it)^r \hat{\psi}(t)$ is the Fourier transform of $(-1)^r [D^r \psi](-x)$, where D is the differential operator with respect to x. Thus, after changing x with $-x$ on both sides of the equation, we find for f the formal expansion

$$f(x) = \exp\left[\sum_{r=1}^{\infty} (\kappa_r - \gamma_r) \frac{(-D)^r}{r!}\right] \psi(x). \tag{3.30}$$

If ψ is chosen as the normal density

$$\phi(x) = \frac{1}{\sqrt{2\pi}\sigma} \exp\left[-\frac{(x-\mu)^2}{2\sigma^2}\right] \tag{3.31}$$

with mean and variance as given by f, that is, mean $\mu = \kappa_1$ and variance $\sigma^2 = \kappa_2$, then the expansion becomes

$$f(x) = \exp\left[\sum_{r=3}^{\infty} \kappa_r \frac{(-D)^r}{r!}\right] \phi(x), \tag{3.32}$$

since $\gamma_r = 0$ for all $r > 2$, as higher cumulants of the normal distribution are 0. By expanding the exponential and collecting terms according to the order of the derivatives, we arrive at the Gram-Charlier A series. Such an expansion can be written compactly in terms of Bell polynomials as

$$\exp\left[\sum_{r=3}^{\infty} \kappa_r \frac{(-D)^r}{r!}\right] = \sum_{n=0}^{\infty} B_n(0, 0, \kappa_3, \ldots, \kappa_n) \frac{(-D)^n}{n!}. \tag{3.33}$$

Since the nth derivative of the Gaussian function ϕ is given in terms of Hermite polynomial as

$$\phi^{(n)}(x) = \frac{(-1)^n}{\sigma^n} He_n\left(\frac{x-\mu}{\sigma}\right) \phi(x), \tag{3.34}$$

this gives us the final expression of the Gram-Charlier A series as

$$f(x) = \phi(x) \sum_{n=0}^{\infty} \frac{1}{n!\sigma^n} B_n(0, 0, \kappa_3, \ldots, \kappa_n) He_n\left(\frac{x-\mu}{\sigma}\right). \tag{3.35}$$

If we include only the first two correction terms to the normal distribution, we obtain

$$f(x) \approx \frac{1}{\sqrt{2\pi}\sigma} \exp\left[-\frac{(x-\mu)^2}{2\sigma^2}\right]\left[1 + \frac{\kappa_3}{3!\sigma^3} He_3\left(\frac{x-\mu}{\sigma}\right) + \frac{\kappa_4}{4!\sigma^4} He_4\left(\frac{x-\mu}{\sigma}\right)\right], \tag{3.36}$$

with $He_3(x) = x^3 - 3x$ and $He_4(x) = x^4 - 6x^2 + 3$.

Note that this expression is not guaranteed to be positive, and is therefore not a valid probability distribution. The Gram-Charlier A series diverges in many cases of interest. It converges only if $f(x)$ falls off faster than $\exp(-(x^2)/4)$ at infinity (Cramer 1957). When it does not converge, the series is also not a true asymptotic expansion, because it is not possible to estimate the error of the expansion.

The Edgeworth Series

Edgeworth developed a similar expansion as an improvement to the central limit theorem. The advantage of the Edgeworth series is that the error is controlled, so that it is a true asymptotic expansion.

Let $\{Z_i\}$ be a sequence of independent and identically distributed random variables with finite meanμ and variance σ^2, and let X_n be their standardized sums:

$$X_n = \frac{1}{\sqrt{n}} \sum_{i=1}^{n} \frac{Z_i - \mu}{\sigma}. \tag{3.37}$$

Let F_n denote the cumulative distribution functions of the variables X_n. Then by the central limit theorem,

$$\lim_{n\to\infty} F_n(x) = \Phi(x) \equiv \int_{-\infty}^{x} \frac{1}{\sqrt{2\pi}} e^{-\frac{1}{2}q^2} dq \tag{3.38}$$

for every x, as long as the mean and variance are finite.

The standardization of $\{Z_i\}$ ensures that the first two cumulants of X_n are $\kappa_1^{F_n} = 0$ and $\kappa_2^{F_n} = 1$. Now assume that, in addition to having mean μ and variance σ^2, the i.i.d. random variables Z_i have higher cumulants κ_r. From the additivity and homogeneity properties of cumulants, the cumulants of X_n in terms of the cumulants of Z_i are for $r \geq 2$,

$$\kappa_r^{F_n} = \frac{n\kappa_r}{\sigma^r n^{r/2}} = \frac{\lambda_r}{n^{r/2-1}} \quad \text{where} \quad \lambda_r = \frac{\kappa_r}{\sigma^r}. \tag{3.39}$$

If we expand the formal expression of the characteristic function $\hat{f}_n(t)$ of F_n in terms of the standard normal distribution, that is, if we set

$$\phi(x) = \frac{1}{\sqrt{2\pi}} \exp\left(-\frac{1}{2}x^2\right), \tag{3.40}$$

then the cumulant differences in the expansion are

$$\begin{aligned} \kappa_1^{F_n} - \gamma_1 &= 0, \\ \kappa_2^{F_n} - \gamma_2 &= 0, \\ \kappa_r^{F_n} - \gamma_r &= \frac{\lambda_r}{n^{r/2-1}}; \qquad r \geq 3. \end{aligned} \tag{3.41}$$

The Gram-Charlier A series for the density function of X_n is now

$$f_n(x) = \phi(x) \sum_{r=0}^{\infty} \frac{1}{r!} B_r\left(0, 0, \frac{\lambda_3}{n^{1/2}}, \ldots, \frac{\lambda_r}{n^{r/2-1}}\right) He_r(x). \tag{3.42}$$

The Edgeworth series is developed similarly to the Gram-Charlier A series, only that now terms are collected according to powers of n. The coefficients of $nm/2$ term can be obtained by collecting the monomials of the Bell polynomials corresponding to the integer partitions of m. Thus, we have the characteristic function as

$$\hat{f}_n(t) = \left[1 + \sum_{j=1}^{\infty} \frac{P_j(it)}{n^{j/2}}\right] \exp(-t^2/2), \tag{3.43}$$

where $P_j(x)$ is a polynomial of degree $3j$. Again, after inverse Fourier transform, the density function f_n follows as

$$f_n(x) = \phi(x) + \sum_{j=1}^{\infty} \frac{P_j(-D)}{n^{j/2}} \phi(x) . \tag{3.44}$$

Likewise, integrating the series, we obtain the distribution function

$$F_n(x) = \Phi(x) + \sum_{j=1}^{\infty} \frac{1}{n^{j/2}} \frac{P_j(-D)}{D} \phi(x) . \tag{3.45}$$

We can explicitly write the polynomial $P_m(-D)$ as

$$P_m(-D) = \sum \prod_i \frac{1}{k_i!} \left(\frac{\lambda_{l_i}}{l_i!} \right)^{k_i} (-D)^s , \tag{3.46}$$

where the summation is over all the integer partitions of m such that $\sum_i ik_i = m$ and $l_i = i + 2$ and $s = \sum_i k_i l_i$.

3.2.1.6 Solving Permutation via Pdfs-Edgeworth/Gram-Charlier

From the above derivation of the two expansions we see that we have an analytic handle to find the best match of the Hermite polynome related source to not fit to the Gram-Charlier expansion. Numerically it would not possible with a limited amount of data to do these statistical derivations with significance, but if we remember that we are completely within an analytical formulation, we can add to the Blind Separation cost function a constraint for solving the permutation.

3.2.2 *Design Space as Tree of Options*

Summarizing how far we got in this Chapter, we have seen that we can find a generic solution for the Random Dynamic System the Field Robot uses to model the situations it is in. Or we can use the same generic solution methodology (because it is generic) for the Designer of the Field Robot to model the situations the Field Robot is in, with the Field Robot internal modeling the situation it is in. Inner and outer look at the situations have the same generic methodology to be described. This is very important, as now the Designer sees the Tree of Options (options generated by the Blind Separation approach) as stochastic bisimulations which all solve the requirement/specification task. A bisimulation is an equivalence relation used in computer science and formal methods to identify pairs of processes with identical behavior, despite potentially different internal structures or implementations. This concept plays a crucial role in verifying and analyzing complex systems by reducing the complexity through abstraction while preserving their essential properties. Bisimulations are particularly useful for model checking, refinement verification, and equivalence checking tasks in various domains such as software engineering, hardware design, and concurrent systems analysis. The Designer can further elaborate on making choices which solution from this tree can be identified to be the best.

At this point in the discussion we need to reflect that we are dealing with complex real situations for field robotics. Options are numerous, and it is well known that it is computationally very expensive to explicitly name/list them all. Hence, the Designer, looking at a very large design space, should avoid to program a robot by listing all options together with situational awareness detection and behavioral solution. On the other hand, even if the Designer would be able to do so, the Field Robot would carry all programs for all options and need to calculate entire parts of decision trees, which is for an online version of an energy efficient Robot not viable. A third issue is intra-design-space emergence that occurs if the robot is implemented and chooses one of the preplanned options, but from this bottom-up (as it is made within a concrete situation) choice-selection situations occur that are missing in the somehow cut (to fit into the Designer's available computational environment) version of elaborated options.

After this discussion, it seems to be wise for the Designer to keep the analytic representation of the Design Space in the form of carrying only implicitly all options, and to design the Field Robot to be prepared to handle all situations on its own, but only being prepared to handle its current situation, not being aware of all options in other situations which are not like the present situation. In other words, on the one hand side, we keep the idea of both Designer and Field Robot to carry a complete Design Space representation in order to model the performance of the autonomous Field Robot in terms of the difficulty to solve the given Design task in general, but on the other hand, we model the Field Robot only as a machine reflecting on its current situation.

From a more abstract point of view, and we will elaborate on this in the following chapters, we are aiming at separating the task's difficulty from the capacity of the autonomous approach used to program the robot.

For now, we need more mathematical tools in order to see how the solutions of the control equations (Stochastic Optimal Control, Koopman operator) discussed in the separation approach look like and to being able to model what the Field Robot can compute when it is on its own (Hidden Markov Model, Integrated Nested Laplace Approximation).

3.2.3 *Analytical Tools*

The Koopman Operator is an extension and generalization of the Fokker-Planck equation, which serves as a powerful tool for analyzing nonlinear dynamical systems in terms of their infinite-dimensional Hilbert space structure. It was introduced by Dutch mathematician Balthasar van der Pol and further developed by Dutch physicist Hendrik Koopman. The primary advantage of the Koopman operator is that it allows us to study complex, nonlinear dynamical systems without resorting to simplifying assumptions or approximations typically required in other methods. In contrast, the Kramers-Moyal (KM) equation, also known as the Fokker-Planck equation, describes the time evolution of a probability distribution function for a system with both deterministic and stochastic components. It is often used to model Brownian motion or other diffusion processes in continuous systems. The KM equation is based on the assumption that the system's dynamics can be described by a set of linear partial differential equations, which may not always hold true for complex nonlinear systems. While both operators are related and have similarities, they serve different purposes within the context of dynamical systems theory. The Koopman operator provides a more general framework that is particularly well-suited to studying nonlinear dynamics without making restrictive assumptions about the system's behavior, whereas the Kramers-Moyal equation offers an approximate description for linearized diffusion processes in continuous systems.

3.2.3.1 Stochastic Optimal Control

Following [14], H^∞ control theory is a branch of control theory that deals with the design of feedback systems such that the closed-loop system has bounded input-output behavior over all possible inputs and disturbances. The main objective of H^∞ control theory is to ensure stability, robustness, and performance optimization for a given dynamic system. H^∞ control theory provides a framework for designing controllers that can guarantee a certain level of performance in terms of the maximum allowable gain between input and output signals. This approach is particularly useful when dealing with systems that are subject to uncertainties or disturbances, as it allows for the design of controllers that can ensure stability even in the presence of these factors. The H^∞ control problem can be formulated as a constrained optimization problem, where the objective is to minimize a performance index while ensuring that certain gain and phase margins are met. This can be achieved by selecting an appropriate controller structure and tuning its parameters to meet the desired specifications. H^∞ control theory has been applied in a wide range of applications, including aerospace systems, robotics, power electronics, and chemical processes. It is an important tool for ensuring that complex feedback systems operate safely and efficiently under various operating conditions.

3.2.3.2 Koopman

Following [13], the Koopman operator can be used to optimize vectors of initial inputs to get given desired expected values of observables. This can be used inside the Field Robot to perform online control by picking the right system component or to set the internal systems parameters. The analytic solution for the control is inserted into the system description equations.

In terms of a more abstract application in the Design Space Evaluation task, with the help of the Koopman operator, it is possible to set up a probabilistic inverse problem, i.e. given expectations of observables of the output, what is a valid input distribution? Translated this into the requirements formulation for an autonomous Field Robot, by solving this probabilistic inverse problem, we link the requirements to the specification in a Design Space Exploration planning modus. If we overlay the cost function for this inverse problem over the optimization problems we are developing so far, we make sure that the given design framework (requirements in Gaussian and specifications in uniform) are met.

3.2.3.3 Hidden Markov Model (HMM)

Following [4], a hidden Markov model (HMM) is a Markov model in which the observations are dependent on a latent (or "hidden") Markov process (referred to as X). An HMM requires that there is an observable process Y whose outcomes depend on the outcomes of X in a known way. Since X cannot be observed directly, the goal is to learn about state of X by observing Y. By definition of being a Markov model, an HMM has an additional requirement that the outcome of Y at time $t = t_0$ must be "influenced" exclusively by the outcome of X at $t = t_0$ and that the outcomes of X and Y at $t < t_0$ must be conditionally independent of Y at $t = t_0$ given X at $t = t_0$. Estimation of the parameters in an HMM can be performed using maximum likelihood. For linear chain HMMs, the Baum-Welch algorithm can be used to estimate the parameters.

Hidden Markov models are known for their applications to thermodynamics, statistical mechanics, physics, chemistry, economics, finance, signal processing, information theory, pattern recognition—such as speech, handwriting, gesture recognition, part-of-speech tagging, musical score following, partial discharges and bioinformatics.

Definition
Let X_n and Y_n be discrete-time stochastic processes and $n \geq 1$. The pair (X_n, Y_n) is a hidden Markov model if

$$\begin{aligned} &X_n \text{is a Markov process whose behavior is not directly observable ("hidden");} \\ &\mathbf{P}\big(Y_n \in A \mid X_1 = x_1, \ldots, X_n = x_n\big) = \mathbf{P}\big(Y_n \in A \mid X_n = x_n\big), \end{aligned} \tag{3.47}$$

for every $n \geq 1$, $x_1, \ldots, x_n$, and every Borel set A.

Let X_t and Y_t be continuous-time stochastic processes. The pair (X_t, Y_t) is a hidden Markov model if

$$X_t \text{is a Markov process whose behavior is not directly observable ("hidden");}$$
$$\mathbf{P}(Y_{t_0} \in A \mid \{X_t \in B_t\}_{t \le t_0}) = \mathbf{P}(Y_{t_0} \in A \mid X_{t_0} \in B_{t_0}), \tag{3.48}$$

for every t_0, every Borel set$A$, and every family of Borel sets $\{B_t\}_{t \le t_0}$.

Terminology
The states of the process X_n (resp. X_t) are called hidden states, and $\mathbf{P}\big(Y_n \in A \mid X_n = x_n\big)$ (resp. $\mathbf{P}\big(Y_t \in A \mid X_t \in B_t\big)$) is called emission probability or output probability.

Probabilistic parameters of a hidden Markov model:

- X—states
- y—possible observations
- a—state transition probabilities
- b—output probabilities

In the standard type of hidden Markov model considered here, the state space of the hidden variables is discrete, while the observations themselves can either be discrete (typically generated from a categorical distribution) or continuous (typically from a Gaussian distribution). The parameters of a hidden Markov model are of two types, transition probabilities and emission probabilities (also known as output probabilities). The transition probabilities control the way the hidden state at time t is chosen given the hidden state at time $t-1$.

The hidden state space is assumed to consist of one of N possible values, modelled as a categorical distribution. (See the section below on extensions for other possibilities.) This means that for each of the N possible states that a hidden variable at time t can be in, there is a transition probability from this state to each of the N possible states of the hidden variable at time $t+1$, for a total of N^2 transition probabilities. Note that the set of transition probabilities for transitions from any given state must sum to 1. Thus, the $N \times N$ matrix of transition probabilities is a Markov matrix. Because any transition probability can be determined once the others are known, there are a total of $N(N-1)$ transition parameters.

In addition, for each of the N possible states, there is a set of emission probabilities governing the distribution of the observed variable at a particular time given the state of the hidden variable at that time. The size of this set depends on the nature of the observed variable. For example, if the observed variable is discrete with M possible values, governed by a categorical distribution, there will be $M-1$ separate parameters, for a total of $N(M-1)$ emission parameters over all hidden states. On the other hand, if the observed variable is an M-dimensional vector distributed according to an arbitrary multivariate Gaussian distribution, there will be M parameters controlling the means and $\frac{M(M+1)}{2}$ parameters controlling the covariance matrix, for a total of $N(M+\frac{M(M+1)}{2}) = \frac{NM(M+3)}{2} = O(NM^2)$ emission parameters. (In such a case, unless the value of M is small, it may be more practical to restrict the nature of the covariances between individual elements of the observation vector, e.g. by assuming that the elements are independent of each other, or less restrictively, are independent of all but a fixed number of adjacent elements.)

Temporal Evolution of a Hidden Markov Model

- Inference: Given that we have observed the output sequence, we may be interested in the most likely sequence of states that could have produced it. We can find the most likely sequence by evaluating the joint probability of both the state sequence and the observations for each case (simply by multiplying the probability values). In general, this type of problem (i.e. finding the most likely explanation for an observation sequence) can be solved efficiently using the Viterbi algorithm.
- Probability of an observed sequence: The task is to compute in a best way, given the parameters of the model, the probability of a particular output sequence. This requires summation over all possible state sequences:
  The probability of observing a sequence

$$Y = y(0), y(1), \dots, y(L-1) \tag{3.49}$$

  of length L is given by

$$P(Y) = \sum_X P(Y \mid X) P(X), \tag{3.50}$$

  where the sum runs over all possible hidden-node sequences

$$X = x(0), x(1), \dots, x(L-1). \tag{3.51}$$

  Applying the principle of dynamic programming, this problem, too, can be handled efficiently using the forward algorithm. Probability of the latent variables
  A number of related tasks ask about the probability of one or more of the latent variables, given the model's parameters and a sequence of observations $y(1), \dots, y(t)$.
- Filtering: The task is to compute, given the model's parameters and a sequence of observations, the distribution over hidden states of the last latent variable at the end of the sequence, i.e. to compute $P(x(t) \mid y(1), \dots, y(t))$. This task is used when the sequence of latent variables is thought of as the underlying states that a process moves through at a sequence of points in time, with corresponding observations at each point. Then, it is natural to ask about the state of the process at the end. This problem can be handled efficiently using the forward algorithm. An example is when the algorithm is applied to a Hidden Markov Network to determine $\mathrm{P}(h_t \mid v_{1:t})$.
- Smoothing: This is similar to filtering but asks about the distribution of a latent variable somewhere in the middle of a sequence, i.e. to compute $P(x(k) \mid y(1), \dots, y(t))$ for some $k < t$. From the perspective described above, this can be thought of as the probability distribution over hidden states for a point in time k in the past, relative to time t.

The forward-backward algorithm is a good method for computing the smoothed values for all hidden state variables.

- Most likely explanation: The task, unlike the previous two, asks about the joint probability of the entire sequence of hidden states that generated a particular sequence of observations. This task is generally applicable when HMMs are applied to different sorts of problems from those for which the tasks of filtering and smoothing are applicable. An example is part-of-speech tagging, where the hidden states represent the underlying parts of speech corresponding to an observed sequence of words. In this case, what is of interest is the entire sequence of parts of speech, rather than simply the part of speech for a single word, as filtering or smoothing would compute.
  This task requires finding a maximum over all possible state sequences, and can be solved efficiently by the Viterbi algorithm.
- Statistical significance: For some of the above problems, it may also be interesting to ask about statistical significance. What is the probability that a sequence drawn from some null distribution will have an HMM probability (in the case of the forward algorithm) or a maximum state sequence probability (in the case of the Viterbi algorithm) at least as large as that of a particular output sequence? When an HMM is used to evaluate the relevance of a hypothesis for a particular output sequence, the statistical significance indicates the false positive rate associated with failing to reject the hypothesis for the output sequence.
- Learning: The parameter learning task in HMMs is to find, given an output sequence or a set of such sequences, the best set of state transition and emission probabilities. The task is usually to derive the maximum likelihood estimate of the parameters of the HMM given the set of output sequences. No tractable algorithm is known for solving this problem exactly, but a local maximum likelihood can be derived efficiently using the Baum-Welch algorithm or the Baldi-Chauvin algorithm. The Baum-Welch algorithm is a special case of the expectation-maximization algorithm.
  If the HMMs are used for time series prediction, more sophisticated Bayesian inference methods, like Markov chain Monte Carlo (MCMC) sampling are proven to be favorable over finding a single maximum likelihood model both in terms of accuracy and stability. Since MCMC imposes significant computational burden, in cases where computational scalability is also of interest, one may alternatively resort to variational approximations to Bayesian inference. Indeed, approximate variational inference offers computational efficiency comparable to expectation-maximization, while yielding an accuracy profile only slightly inferior to exact MCMC-type Bayesian inference.

3.2.3.4 Integrated Nested Laplace Approximation (INLA)

Following [5], the Integrated Nested Laplace Approximation (INLA) approach is a computational approach to perform fast Bayesian inference in the broad class of latent Gaussian models, that is, models of an outcome variable y_i that assume independence conditional on some underlying (unknown) latent field and a vector of hyperparameters μ. It was proposed as an alternative to the usually time consuming MCMC methods. Unlike MCMC where posterior inference is sample-based, the INLA computational approach directly approximates the posteriors of interest with a closed form expression. Therefore, problems of convergence and mixing, inherent to MCMC runs, are not an issue. The main aim of the INLA approach is to approximate the marginal posteriors for the latent variables as well as for the hyperparameters of the Gaussian latent mode.

We have already encountered the usefulness of incorporating a latent layer in the formulation of the Bayesian mixture approach for classification in the Signals chapter. Furthermore, we have seen in the previous chapter on HMMs that MCMC sampling is involved to find solutions for the HMM-associated problem the Field Robot has. Since we are interested in a completely analytical description and solution of the Design Space Exploration, the INLA approach is a suitable replacement of the MCMC algorithms.

INLA is based on an efficient combination of (analytical Gaussian) Laplace approximations to full conditionals and numerical integration routines to integrate out the hyperparameters μ, where we in Design Space Exploration can use analytic integration because we are working with Gaussian distributions which allow such an analytic treatment.

The INLA framework can be described as follows. First of all, $y = (y_1, \ldots, y_n)$ is a vector of observed variables whose distribution is in the exponential family (in most cases), and the mean μ_i (for observation y_i) is conveniently linked to the linear predictor η_i using an appropriate link function (it is also possible to link the predictor to e.g. a quantile). The linear predictor can include terms on covariates (i.e., fixed effects) and different types of random effects. The vector of all latent effects will be denoted by x, and it will include the linear predictor, coefficients of the covariates, etc. In addition, the distribution of y will likely depend on some vector of hyperparameters θ. The distribution of the vector of latent effects x is assumed to be Gaussian Markov random field (GMRF). This GMRF will have a zero mean and precision matrix $Q(\theta_2)$, with θ_2 a vector of hyperparameters. The vector of all hyperparameters in the model will be denoted by $\theta = (\theta_1, \theta_2)$. Furthermore, observations are assumed to be independent given the vector of latent effects and the hyperparameters. This means that the likelihood can be written as

$$\pi(y|x, \theta) = \prod_{i \in I} \pi(y_i|\eta_i, \theta). \tag{3.52}$$

Here, η_i is the latent linear predictor (which is part of the vector x of latent effects) and set I contains indices for all observed values of y. Some of the values may not have been observed.

The aim of the INLA methodology is to approximate the posterior marginals of the model effects and hyperparameters. This is achieved by exploiting the computational properties of GMRF and the Laplace approximation for multidimensional integration.

In the Design Space Exploration approach further outlined in the next chapters, we need to model how the Field Robot performs its calculations and which results it can achieve. The INLA formulas replace the HMM formulations for the model of the decision making inside the Field Robot in a completely analytical manner. Again, large formulas that we can insert into the framework we have outlined with the Cascaded Gaussian Functional, which will be further elaborated on in the next chapters.

Take-Away for Design Space Exploration
The more physical insight we can incorporate into the models the Field Robot is using, the better its synchronization with the real world will be. However, if we simply aim to put all our insight into a robot, it cannot calculate this world formula. Decisions have to be made: What is relevant and what is not relevant for the given task. This decision has to be made within the robot at the current situation. This avoids the need to write a program code for the Field Robot containing the world formula. On the the other hand, does the designer have a world formula? Can the designer predict all situations? What about situations that emerge due to the embodyment of the designed Field Robot? Taking these two perspectives together, we should not aim to implement a world formula into the robot, but rather implement a mechanism which allows the Field Robot to calculate a local and myopic world around it whereby this local world model is then updated by the robot as fast as necessary to fulfill the given task. Design Space Exploration is then about finding a formulation for this mechanism and about identifying where this mechanism leads to effective, robust, and efficient fulfilments of the given task.

References

1. Ehlers, F.: Controller and observer design for many-robot surveillance systems utilizing non-linear blind source separation. In: Many-Robot Systems: Crossing the Reality Gap, St. Paul, USA (2012)
2. Ehlers, F.: Einrichtung zur adaptiven Entkoppelung a-priori unbekannter dynamischer Systeme, Germany, applied (2000)
3. Preston, L., Paracousti-U.Q.: A Stochastic 3-D Acoustic Wave Propagation Algorithm, SAND2017-10143, Sandia National Laboratories, Albuquerque, NM (2017)
4. Hidden Markov Model. https://en.m.wikipedia.org/wiki/Hidden_Markov_model. Cited 2024/08/12
5. INLA, https://www.r-inla.org/. Cited 2024/07/29
6. Polynomial Chaos Expansion. https://github.com/timueh/PolyChaos.jl. Cited 2024/07/25
7. Edgeworth Expansion. https://en.wikipedia.org/wiki/Edgeworth_series. Cited 2024/07/27
8. Honerkamp, J.: Stochastische Dynamische Systeme. VCH (1990)
9. Poor, H.V.: An Introduction to Signal Detection and Estimation. Springer (1994)
10. Kappen, H.J.: Path integrals and symmetry breaking for optimal control theory. J. Stat. Mech. 11011 (2005)

11. McDonald, M., Balaji, B.: Track-before-detect using swerling 0, 1 and 3 target models for small manoeuvring maritime targets. EURASIP J. Adv. Signal Process. (2008)
12. Xing, J., Wang, H., Oster, G.: From continuum Fokker-Planck models to discrete kinetic models. Biophys. J., v **89**, 1551–1563 (2005). Sept
13. Meyers, J., Rogers, J., Gerlach, A.: Koopman operator method for solution of generalized aggregate data inverse problems. J. Comp. Phys, **428**, (2021)
14. Charalambous, C. D.: Stochastic nonlinear minimax dynamic games with noisy measurements. IEEE Trans. Automatic Control **48**(2) (2003)
15. Lefebvre, T.: On moment estimation from polynomial chaos expansion models. IEEE Control Syst. Lett. **5**(5), 1519–1524 (2021). https://doi.org/10.1109/LCSYS.2020.3040851. Nov
16. Han, X., Kloeden, P.E.: Random ordinary differential equations and their numerical solution. Probability Theory and Stochastic Modelling 85. https://doi.org/10.1007/978-981-10-6265-0_18

Chapter 4
Architectures

Abstract In a way the mathematical models inside the robot make it possible that the robot simulates the real world. This simulation is done with a network of equations, rather complex sets of equations with a lot of parameters. One could call these equations Functional-Mock-Ups, and the simulated Robot could be named a Digital Twin under certain circumstance, i.e. when the Digital Twin and the Robot are synchronized, making sure that simulation and reality do not differ too much (with a distance measure to be set by a user of both, the digital twin and the robot system). If this simulation is then used to check different options for the best next action of the robot, then the model knowledge is incorporated into the robot. If the simulation of the robot automatically choosing the best next action is used to optimize the robot design, the architecture can also be used to setup such extended simulations. (However, optimizing optimized robots with a modelled synchronization mechanism is a heavy workload for the simulating computer. But, at least, the simulation setup can be described.) Instead of performing simulations, we are able to solve the optimization equations analytically, as we are able to generate the Cascaded Gaussian Functionals as explained in the previous chapter. Hence, we are able to provide an automatic software generation from the simulation setup.

Top-Down View

In the last two chapters, we investigated the real world from a syntactic (interface) and a semantic (axiom) description view point. We have seen that mathematical equations are used to model the real world inside the robot and that sensor data has to be connected to these models. The models and the connection between them through the fact that the robot semantically interacts with the real world are leading to a complex network of information flow, for each system we have in the real world. The idea is to make an abstraction: Do we have similar connectivity between a robot that does surveillance in space and a robot that does surveillance in air? The abstraction can go further: Are the system design, the system maintenance and the system simulation similar?

F. Ehlers, *Design Space Exploration in Robotics*, Studies in Systems, Decision and Control 229, https://doi.org/10.1007/978-3-031-78740-9_4

The organisation of these similarities in an architectural framework helps to understand the systems/compare the systems/describe the systems/talk about the systems, and finally to automatically generate software for the systems. What we are looking for is a reference view, i.e. a mathematically abstracted approach to generate ontology elements which can be syntactically elaborated to provide code inside the (to be designed) robot. Again, if an approach can be used inside the robot, the same approach can be used by the Designer, and vice versa. In other words, with the organisation of system details in an architectural environment, we enable both, an automated code generation inside the robot, and an automated code generation for the designer. The key for both is to connect the hierarchical set of equations and the tree of options developed in the previous chapter to the best-practices and standards of existing architecture and system-of-systems modelling approaches.

Coming from the first part of the Introduction, now that we can interpret signals and apply a control to get to our operational goal, we can step one level of abstraction upwards: We have seen that things match with services they are either part of or get manipulated by them. If the outer world has some structure, this thing-to-service abstraction seems to be a common feature. The idea is to describe the outer world by a service-oriented viewpoint, adding categories related to them and arrive at architectures. From a best practice outcome of a longer history of unifying approaches to procure/sell system that can apply a specific service, or platforms reaching a given goal, we introduce the Unified Architecture Framework (UAF), noting that there is a built-in path to combining architectures, letting the work with the UAF be described separately in a top-down and a bottom-up direction. For the top-down-direction: If a system is understood as a system that comprises of other systems, then the interaction between the sub-systems can be described with the help of the SysML, even one level more down with SysPhS as a link to the physical outer world. As a standard SysPhS provides entries into many physical thing and service descriptions, providing unified/standard names and units. For the bottom-up direction, an existing part (or building block) is interpreted as a little Enterprise: Incoming information is sold to a controller as customer, and the money earned is related to how much you cover the given operational goal. Each starting building block is considered to be a micro-service, and the next level of enterprise architecture is reached by combining the existing services with the help of a new business process. Where we end up is the digital description of a dynamic system, a digital twin, which can be inserted into Dynamic Systems already described by the SysPhS in the top-down approach. This confrontation of the two approaches is of essential importance: In Chap. 6 we will see how far we can get with the assumption of a full overlap. If the overlap is 100%, we call the system technically engineered. As we will find out in Chap. 7, relaxing the assumption that top-down and bottom-up have to match 100% avoids criticality, but on the other hand introduces inconsistencies.

In the next Sections, we describe potential interfaces between UAF and SysML, specifically already for the physical domain by focusing on SysPhS. We refer to the Semantic Sensor Network as a description of a physical real-world experiment, and to the Platform Stack Architectural Framework of the Digital Twin as a simulation of the same physical experiment. We will identify that the interfaces between NAF and SysML can be linked by connecting the appropriate services in both, the Semantic Sensor Network as a description of a physical experiment and the Platform Stack Architectural Framework for the Digital Twin as a simulation of the physical experiment plus synchronization and dashboard services. Since we are in the design phase for our system, these services need to be generated first. The various connection possibilities represent service options and their interface environment represent the generation constraints for the Artificial Intelligence Functionalities in the Robotic Applications. Behavioral Robotics provides a seamless link from operational procedures formulated in the UAF via SysPhS SysML-constructions to the programming framework of artificial intelligence functions onboard autonomous platforms, and automated software generation can generate the algorithm to be implemented as a Service into an Internet-of-Things driven semantic sensor network or Digital Twin cloud framework. From the game-theoretic viewpoint of balancing between demands and realistically implementable software abilities these constraints are already available by the tree of options for the polynomials we have developed in the previous chapter. Taking their generic coefficients to be parameters of an optimization process related to the design process, a generic mathematical technique suitable for "automated software generation accompanied by a computationally viable verification and validation process" approach becomes algorithmically possible. In other words, available best-practices and standards are capable to cover the mathematical framework developed in the previous chapter. Algorithmially pressing a bottom for starting optimization in the Design Space is a viable approach, and in the light of the system-of-system complexity it seems to be the only viable approach to use an overarching abstract tool like Enterprise Architectures to tackle this complexity. The important point of this chapter is that the complexity of Design Space Exploration in Robotics can be captured and that by this capturing a linkage between low-level simulations and abstract Enterprise Architectures becomes possible. This is an important new look on Enterprise Architectures because normally they are just seen as a tool for the work in the abstract domains, leaving the detailed simulations domains to other algorithmic simulation tools.

4.1 System Model

A System Model describes the Structure, Behavior, Requirements, and Parametrics together with the connections between these blocks. As an Integration Framework the System Model serves as interface to External Requirements (traceability rationale), Systems Specifications (viewpoints), Analysis Models (analysis needs and performance estimates), and to various Design Models (Mechanical, Electrical, Software,

Testing) which form the "system framework for design"-interface. Utilizing SysML and its extension SysPhS for data gathering and platform actuation, it is possible to describe in details the system components and its interfaces to the environment, already aiming at interfacing to major Modelling and Simulation software packages. Via UAF, this interface from classical physics to software tools is available for applications in any domain. Whilst UAF is an abstraction of a categorical foundational ontology [12], the origin of SysML as an UML derivate can be interpreted as a specification of a generic language. In this Section, we describe potential interfaces between NAF and SysML, specifically already for the physical domain by focussing on SysPhS.

4.2 Enterprise Architectures

An enterprise architecture (EA) is a comprehensive framework that provides a holistic view of an organization's business processes, information technology systems, and infrastructure components. It serves as the foundation for aligning strategic goals with operational capabilities to achieve optimal performance, efficiency, and adaptability in today's dynamic market landscape. The primary objective of EA is to ensure effective communication among various stakeholders within an enterprise while enabling seamless collaboration between business units, IT departments, and other relevant parties. An EA can be applied across a wide range of industries and organizational sizes, from small businesses to large multinational corporations. By defining the structure, governance, and management processes for information systems and technology resources, an enterprise architecture facilitates strategic decision-making, risk mitigation, and continuous improvement initiatives throughout the organization's lifecycle. Some key components of a successful EA include:

- Business Architecture—Defines the organizational structure, strategies, goals, and processes to achieve desired business outcomes.
- Information Systems (IS) Architecture—Focuses on information systems, data management, and integration across various departments within an organization.
- Infrastructure Architecture—Addresses the physical and virtual infrastructure components required for efficient operation of IT systems and resources.
- Application Architecture—Outlines the design principles, patterns, and best practices for developing and maintaining software applications in alignment with business objectives.
- Security and Compliance—Ensures that all aspects of an organization's EA adhere to relevant security standards, regulations, and industry best practices.
- Governance and Management—Establishes a clear framework for decision-making, resource allocation, and performance measurement within the context of an enterprise architecture.

In summary, an enterprise architecture is essential for enterprises seeking to optimize their business processes, IT systems, and infrastructure components while ensuring effective communication among stakeholders and aligning strategic goals with operational capabilities.

4.3 Unified Architecture Framework (UAF)

For the model-based approach, we rely on the Unified Architecture Framework (UAF) [12] as a technical foundation and on mathematical frameworks dealing with uncertainty. The UAF provides an architectural modeling framework for mapping structures, behavior as well as parameters and requirements using well defined viewpoints for different concerns such as strategic, operational, services and resources. Here, in terms of modeling, the UAF provides a metamodel, which is based on the SysML, which is well established for model-based systems engineering, but needs to be extended to incorporate the mathematical frameworks dealing with uncertainty. Accordingly, the UAF enables storage, maintenance, and visualization of the details of complex Systems-of-Systems in terms of diagrams following a well-defined notation.

4.4 System-Of-Systems (SoS)

Internet-of-Things driven Semantic Sensor Networks as one example for the suitability of the Semantic Sensor Network Ontology, (see especially Fig. 10 in [6] for classes and relationships involved in Observation (SOSA/SSN)) are mentioned. The example given in Sect. 4.5 elaborates on the "detects" relationship.

4.5 System Modeling Language (SysML)

There are two versions available for the System Modeling Language:

- SysML 1.x. Assuming that the input-output-relationship of a deterministic System-of-Systems can be described by interfacing descriptions of state space interactions as differential equations and descriptions of state machines, a simulation of the System-of-Systems can start with given initial conditions and shows the output corresponding to input signals as a causal formulation. Alternatively, assuming that the System-of-Systems is reacting to given external input, which can be thought of as a constraint at some descriptive states of the system, the System-of-Systems is simulated in an "a-causal" manner by letting the system run under these given constraints. SysML provides the syntactical language elements to set up both ways (causal and a-causal) of simulating the system.

- SysML 2.0. [4] The SysML environment shows up as a very "dynamical" standardization environment. This is due to the introduction of SysML 2.0. SysML 2.0 has more distance to UML than SysML 1.0, maybe indicating the necessity to position the SysML as standard quasi between UML and UAF, in order to become most useful. As an example for the suitability of using SysML 2.0 in the application domain of artificial intelligence functionalities, the definition of the FailureModeling is highlighted, whereby the attribute probability links directly to engineering terms of Probability of Detection (PD) and Probability of False Alarms (PFa) which have been coming up in the Decision-Making Section of Chap. 2.

Failure Modeling

```
library package FailureModeling{

  abstract occurrence def Situation;

  abstract occurrence situations : Situation[*] nonunique;

  abstract occurrence def Cause {

    attribute probability : Real;

  }

  ...

}
```

A second example is the downscaling process for requirements which is supported in SysML 2.0, which fits to the "automated software generation accompanied by a computationally viable verification and validation process".

4.6 SysML Physical Interaction and Signal Flow Simulation (SysPhS)

Following [5], with the SysML-Extension for Physical Interaction and Signal Flow Simulation (SysPhS) a standard description of a physical layer (dealing with classical physics described by charge, volume, angular momentum, linear momentum, and entropy) for data gathering and platform actuation is available, already aiming at interfacing to major Modelling and Simulation software packages. Hence, SysPhS is a platform-independent extension of SysML for physical interaction and signal flow (also known as lumped parameter, one-dimensional, or network models, appearing in simulation platforms supporting the Modelica language and in MathWorks Simulink and Simscape).

4.7 Business Process

4.7.1 BPMN versus CMMN

Business Process Model and Notation (BPMN) is a standard for business process modeling that provides a graphical notation for specifying business processes in a Business Process Diagram (BPD), based on a flowcharting technique very similar to activity diagrams from Unified Modeling Language (UML). The objective of BPMN is to support business process management, for both technical users and business users, by providing a notation that is intuitive to business users, yet able to represent complex process semantics. The BPMN specification also provides a mapping between the graphics of the notation and the underlying constructs of execution languages, particularly Business Process Execution Language (BPEL). BPMN has been designed to provide a standard notation readily understandable by all business stakeholders, typically including business analysts, technical developers and business managers. BPMN can therefore be used to support the generally desirable aim of all stakeholders on a project adopting a common language to describe processes, helping to avoid communication gaps that can arise between business process design and implementation. BPMN is one of a number of business process modeling language standards used by modeling tools and processes. While the current variety of languages may suit different modeling environments, there are those who advocate for the development or emergence of a single, comprehensive standard, combining the strengths of different existing languages. It is suggested that in time, this could help to unify the expression of basic business process concepts (e.g., public and private processes, choreographies), as well as advanced process concepts (e.g., exception handling, transaction compensation).

Case Management Model and Notation (CMMN) is a graphical representation for graphically expressing a Case, as well as an interchange format for exchanging Case models among different tools. In this context: "A Case is a proceeding that involves actions taken regarding a subject in a particular situation to achieve a desired outcome" and it is derived from the concept of Case management used e.g. in the legal and medical worlds.

While BPMN deals with predictable processes, CMMN is more oriented to focus on supporting processes that are intrinsically unpredictable, not repeatable, weakly-structured, and knowledge-intensive. CMMN can often require ad-hoc actions.

4.7.2 DMN

In business analysis, the Decision Model and Notation (DMN) is a standard published by the Object Management Group that provides a standard approach for describing and modeling repeatable decisions within organizations. This ensures that decision models are interchangeable across different organizations. The DMN notation is designed to be easily understood by both business users and IT professionals, facilitating collaboration among various groups involved in defining a decision model:

1. Business people who manage and monitor the decisions
2. Business analysts or functional analysts responsible for documenting initial decision requirements and specifying detailed decision models and decision logic
3. Technical developers responsible for automating systems that make these decisions

The DMN standard can be used independently but is also complementary to other standards like BPMN (Business Process Model and Notation) and CMMN (Case Management Model and Notation). In BPMN, a special kind of activity called the Business Rule Task has been defined. This task provides a mechanism for processes to provide input to a business rule engine and get output from calculations that the business rule engine might provide. DMN is now recognized as a standard for Business Analysis according to BABOK v3 (Business Analyst Body of Knowledge).

The integration between DMN and BPMN has been studied extensively, and they are designed to work together. Decisions in DMN can be explicitly linked to processes and tasks that use these decisions. By moving process logic into decision services using DMN, business process models can be simplified. A stateless, side-effect free Decision Service is expected to be deployed from a decision made in DMN and invoked from the business process. This allows data within the process to be mapped to inputs and outputs of the decision service.

4.7.3 Syntax to Semantics

Syntactically describing a business process is one part of the story, connecting this syntax to the semantics of data having received (or to be simulated/predicted) is another part of the story of digitalisation. Implementing this connectivity in a generic (and standardized) manner is an important step to make the Digital Twin concept work. In this Subsection, we mention the bupaR approach and the XES standard for data exchange.

4.7.3.1 bupaR

Following [10], we introduce important terminology:

- Logs: Transforming your raw data into an event log object is one of the most challenging tasks in process analysis.
- Case: The subject of your process, e.g. a customer, an order, a patient.
- Activity: A step in your process, e.g. receive order, sent payment, perform MRI SCAN, etc.
- Activity instance: The execution of a specific step for a specific case.
- Event: A registration connected to an activity instance, characterized by a single timestamp. E.g. the start of Perform MRI SCAN for Patient X.
- Resource: A person or machine that is related to the execution of (part of) an activity instance. E.g. the radiologist in charge of our MRI SCAN.
- Lifecycle status: An indication of the status of an activity instance connect to an event. Typical values are start, complete. Other possible values are schedule, suspend, resume, etc.
- Trace: A sequence of activities. The activity instances that belong to a case will result to a specific trace when ordered by the time each instance occurred.

bupaR supports two different kinds of log formats, both of which are an extension on R data.frame:

- eventlog: Event logs are created from data.frame in which each row represents a single event. This means that it has a single timestamp.
- activitylog: Activity logs are created from data.frame in which each row represents a single activity instances. This means it can has multiple timestamps, stored in different columns.

The data model below shows the difference between these two levels of observations, i.e. activity instances versus events.

4.7.3.2 XES-Standard

The XES (eXtensible Event Stream) standard defines a grammar for a tag-based language with the aim of providing designers of information systems with a unified and extensible methodology for capturing systems' behaviors by means of event logs and event streams, as defined in the XES standard. This includes an "XML Schema" describing the structure of an XES event log/stream and another "XML Schema" describing the structure of an extension to such a log/stream. Additionally, it contains a basic collection of so-called "XES extensions" that provide semantics to certain attributes as recorded in the event log/stream [9].

The scope of this standard is to define World Wide Web Consortium (W3C) Extensible Markup Language (XML) structure and constraints on the contents of XML 1.1 documents, which can be used to represent extensible event stream (XES) instances. A XES instance corresponds to a file-based event log or a formatted event stream that can be utilized for transferring event-driven data in a unified and extensible manner from one site to another. Typically, the first site generates this event-driven data (e.g., workflow systems, case handling systems, procurement systems, devices like wafer steppers and X-ray machines, or hospitals), while the second site analyzes it (e.g., by data scientists and/or advanced software systems). To transfer event-driven data in a unified manner, this standard includes a W3C XML Schema describing the structure of a XES instance. To transfer this data in an extensible manner, this standard also includes a W3C XML Schema describing the structure of an extension to such a XES instance. Essentially, these extensions provide semantics to the structure as prescribed by the XES instance. Finally, this standard comprises a basic collection of such extensions. The purpose of this standard is to establish a generally acknowledged XML format for exchanging event data between information systems in various application domains and analysis tools for that data. By adhering to the syntax and semantics defined by this standard when transferring event data, its meaning will be clear and well-understood at both sites involved: the site generating the data and the site analyzing it.

4.8 Functional Mock-Up Unit (FMU)

In the architectural domain, an a-causal model description of a system as a Differential-Algebraic-Equation (DAE) is used as a mathematical representation to initiate a simulation of the model [8].

In an implicit or explicit form the DAE is capable to handle continuous and time-discrete variables. The setup fits very well to the Random Dynamical System description in Chap. 3, and to the modelling of the internal behavioral aspects of a Field Robot, because the DAE can serve as a Functional Mock-Up, a module-wise description of the robot, that is available as a standard XML interface between various program packages in the Modelling and Simulation community.

Although the development in this Book is meant to be analytical, the standardization from FMUs can be a good starting point for generating exchangable reference-blocks, filled with mathematical formulas, describing calculation tasks or already generated Design Space Exploration solutions [7].

4.9 Synchronization between Digital Twin and Dynamic System

Whilst the a-causal modelling in the previous Section is useful for the evaluation of system behavior, a causal description with explicit input-output relationships (in simulation languages like simulink) is useful when comparing simulated values with real measured values of a dynamic system. Such synchronization is a core necessity described in the Digital Twin Platform Stack Architectural Framework [1], which has been abstracted from the currently available cloud realizations of Digital Twins. The Framework has been developed by the Digital Twin Consortium and it is highlighting also the various relationships between Application, Integration, Virtual Representation, and the Real World. Security and Trustworthiness are also an important aspect in this Framework, and from the discussion of the Introduction in this Book, the ethical aspects are entering naturally through the necessity to perform validation and verification of the Robots with Artificial Intelligence Functionalities. As the Automated Software Generation described in the next Section is accompanied by Verification and Validation processes, this Book might contribute to possible new ways, other than the currently available solutions, for realizing the various relationships between Application, Integration, Virtual Representation, and the Real World.

4.10 Automated Software Generation

In general, we refer to the idea of generic programming, which is also the idea behind the development of concepts in C++, aiming at full generality/expressiveness, zero overhead compared to hand coding, and consistent well-specified interfaces, whereby all three aims turn out to be not doable in the pure axiomatic digital manner of Turing completeness [2]. In this Section, we will identify that the interfaces between NAF and SysML can be linked by connecting the appropriate services in both service-oriented representations, the Semantic Sensor Network as a description of a physical experiment and the Platform Stack Architectural Framework for the Digital Twin as a simulation plus synchronization to the real data of the physical experiment, whereby the Digital Twin dashboard design can be interpreted as a selection of descriptive service states following a ranking by the relevancy of involved parameters. Since we are in the design phase for our system, no system is available to provide real data. Hence, in order to decide how a service should look like, all options for services as outlined in the tree of options in the previous chapter have to be taken in consideration: The various connection possibilities represent these service options. However, each option has its own interface environment which represents the generation constraints for the Artificial Intelligence Functionalities in the Robotic Applications. These interface environments are described by the Services for Behavioral Robotics. By re-modelling (from the game-theoretic viewpoint of balancing between demands and realistically implementable software abilities) these generation constraints with the help of polynomials and taking their coefficients to be parameters of an optimization process related to the design process, a generic mathematical technique suitable for "automated software generation accompanied by a computationally viable verification and validation process" approach is possible.

4.10.1 Main Parts of the Overall Procedure

In order to build-up a procedural description of our approach, the steps to be performed are summarized in the following pseudo-code:

Program Code

```
(A) loading the options for services and their combinations

(B) evaluating the specific interface environment for each
possible service to mathematically formulate constraints
for an optimization process associated to the
Automated Software Generation of
Artificial Intelligence Functionalities

(C) optimization of a potential which aims at reaching a
maximal independency of demands and abilities given
the constraints formulated in (B).
```

An Artificial Intelligence Functionality in a robotic application can mathematically be described as a generic function $f(x)$ as Taylor series with coefficients a_i. The generation of the Artificial Intelligence Functionality in a robotic application can then be transferred to finding the appropriate coefficients a_i such that the Functionality with input x generates the desired output $f(x)$. The Verification and Validation process checks both: the appropriateness of the coefficients and the reachability of the desired output. Extending the inputs to the Functionality for the inclusion of stochastic processes, the state vector x is written as a combination of deterministic and stochastic components. We restrict our discussion to Gaussian and Uniform stochastic processes, as in the previous chapter, and identify the formulation of signal processing and control algorithms to be inherently Gaussian driven, whilst the requirement formulations are to be effective in an interval. For the Verification and Validation process this means, that on Gaussian input a result according to uniform expectations has to be checked. Furthermore, since we are in the design phase (we do not have real data yet), the evaluation comes with synthetic data. We choose Polynomial Chaos Expansions. These surrogates can also be used to enter an analytic evaluation. This is further elaborated in the following Subsection.

4.10.2 *Functionality Structure and Functionality Parameter Fit*

A top-down processing of all potential solutions is performed, "inventing Autonomous Functionalities" for each Layer with Uniform demands and Gaussian driving processes, i.e. the structures of the mathematical formula are found, leaving adjustment parameters open. A holistic optimization of all potential solutions is performed, i.e. the adjustment parameters are fitted according to the requirements. The optimization process is the non-cooperative game in which the Designer (as player 1) and the AI-generation (as player 2) have to find the equilibrium.

4.10.3 *Filter Design*

As an example, we link to the Decision Making section in Chap. 2, and add the architectural terminology to emphase the possible linkage of mathematics and algorithmic formulation.

A finite state machine is formulated with two states: transponder signal detected versus no transponder signal detected.

SysML 2.0: Statistical treatment with probability of detection (PD) and false alarms (FA) via the FailureModel. Regarding the downscaling process for requirements, in this simple example we only deal with one level here, but more complex systems naturally consist of a hierarchy.

SysPhS: Filter Design as Taylor series with to be adjusted coefficients a_i, whereby the signal from sensing is measured in Volt and calibration terms relative to the pressure values regarding the measurement domain's propagation model are inserted to analyze numerical values.

The ontological skeleton of the detection service has to fit with the ontology of the Internet-of-Things driven semantic sensor network (Semantic Sensor Network), i.e. the "Detect"-service formulation in the Semantic Sensor Network has to be consistent with the SysML 1.x formulation. Similarly, signal generation and propagation models have to be checked for consistence.

For the Digital Twin Platform Stack Architectural Framework (Digital Twin), the simulation model has to be consistent with the SysPhS formulation, including calibration and noise process modelling.

With a consistent preparation of the Matched Filter for the transponder signal and adjustment of the threshold parameter according to given performance demands we are led to a specific point on the Receivers Operating Characteristic (ROC).

Take-Away for Design Space Exploration

The Architecture can be seen as a bookkeeping device for all the details of a Field Robot, utilizing the available best-practices and standards in the system-of-systems design community. The mathematics developed in the previous chapters fit seemlessly to these algorithmic technique of model-based code generation. Hence, as a starting point for the discussions in the next chapter, we can state an, in principle, similarity between Designer and Robot Field Robot for both, mathematics and algorithms.

References

1. McKee, D.: Platform Stack Architectural Framework: An Introductory Guide, A Digital Twin Consortium White Paper, 2023-07-11
2. Stroustrup, B., Concepts: The Future of Generic Programming. https://www.stroustrup.com/good_concepts.pdf. Cited 2024/07/21
3. Ehlers, F.: Acoustic modelling knowledge for design and specification of situation-adaptive distributed networked systems. In Underwater Acoustics Conference and Exhibition (UACE 2019), July 2019
4. Introduction to the SysML v2 Language Textual Notation, Release: 2023-02, https://github.com/Systems-Modeling/SysML-v2-Release/blob/master/doc/Intro%20to%20the%20SysML%20v2%20Language-Textual%20Notation.pdf. Cited 2023/07/24
5. SysML Physical Interaction and Signal Flow Simulation (SysPhS), https://www.omgwiki.org/OMGSysML/doku.php?id=sysml-sysphs:sysml_extension_for_physical_interaction_and_signal_flow_simulation_sysphs. Cited 2023/07/24
6. Semantic Sensor Network Ontology, https://w3c.github.io/sdw/ssn/. Cited 2024/07/22
7. fmi, https://fmi-standard.org/. Cited 2024/07/20
8. The Functional Mockup Interface for Tool independent Exchange of Simulation Models, https://www.startpage.com/do/search?query=Modelica2011%2FThe_Functional_Mockup_Interface.pdf. Cited 2024/07/24
9. IEEE 1849-2023 XES Standard, https://www.xes-standard.org/. Cited 2024/08/24
10. bupar, https://bupaverse.github.io/docs/. Cited 2024/09/24
11. Holt, J., Perry, S.: SysML for System Engineering, IET Books, 2020
12. OMG. Unified Architecture Framework (UAF), Version 1.1, 2020
13. Ascher, D., Ehlers, F.: Seamless Integration of ATR Services into Situation-Adaptive Distributed Systems via Reference Modeling. Proc. Mtgs. Acoust 20 June 2021; 44 (1): 070020. https://doi.org/10.1121/2.0001487

Chapter 5
Robots

Abstract The Field Robot has its model in the architectural framework, and the Field Robot has its own built-in architectural framework that it uses to do Design Space Exploration on the task given to it. In this Chapter, we look for a compatibility between the Field Robot's architectural framework and the Designer's architectural framework, and we look for an automatic way, to let the Field Robot execute its autonomous activities, whereby the Designer needs to be able to model the way the Field Robot is executing these autonomous activities.

Top-Down View
This Chapter marks a central point in the Book's discussions: Whilst in the previous chapters we have not looked differently on the two tasks of making decisions inside a robot and making decisions as a designer of the robot (i.e. the automated code generation method could in principle be running both inside and outside the robot), in this Chapter we discuss view points for the two tasks. Real signals will only be received by the robot in the field, but only one situation will occur at the same time for the robot, and probably the robot will not have to stand all possible situations. In the Designer's world, no real signals are available and the robot has to be prepared to run in all possible situations. As we can imagine from the outline of the calculations to be made for Design Space Explorations, energy and processing constraints will most likely not allow to carry this computational power with it. The robot in the field should be prepared for everything, but efficiently handling only a necessary selection of options, whereby these options might be something that in the Designer's world is only implicitly known (i.e. an emergent feature that can be explained within the modelling equations).

Coming along the story-telling first part of the Introduction, we discuss in Chap. 5 how the "languages" in Architectures can be used to describe our Robot in action, how

F. Ehlers, *Design Space Exploration in Robotics*, Studies in Systems, Decision and Control 229, https://doi.org/10.1007/978-3-031-78740-9_5

the signals it receives can be used to generate a physical model of the environment it is operating in, leading to the finding that it should be possible to describe a Design Space Exploration-Machine, i.e. an algorithm that optimizes design and operational parameters given a task. This Machine would make our Robot become an Artificially Intelligent Robot. A preliminary for reaching this programming goal is to be able to set the three languages (Service-Based Architectures, Signals, Physics) in action all together to allow us to predict the performance of a robot on a specific task. The prediction is meant to be comparable to other robots for the same task. Such performance metric is the key to apply Artificial Intelligence algorithms in Robotics: Optimization algorithms can work on controlling to make the task easier and/or to make the performance of a specific robot better.

5.1 Architectural View

For architectural modeling of a robot, we distinguish between reference modeling (as an autonomous approach to several tasks) as well as domain-specific modeling (as an autonomous approach to a specific task), where the latter describes the application of the reference model to a particular domain for specific scenarios and their requirements resulting in a model configuration. Subsequently, we shortly describe the elements utilized for this application, where our approach is comprised of the following elements: [5]

- Viewpoints are defined by their subjects of concerns such as Strategic, Operational, Service or Resources as well as their aspects of concerns, such as Structure, Behaviors, Parameters or Requirements. Here, a reference model for the robot is composed of layers describing different aspects and subjects of concern, which are defined using specific viewpoints. In the ISO/IEC/IEEE 42020, a viewpoint represents the "architecture conventions for the construction, interpretation and use of architecture views to address specific concerns about the architecture entity", where a concern relates to a "matter of interest or importance to a stakeholder".
- Views describe individual perspectives on the model itself. They are constructed based on viewpoints and use assets. According to the ISO/IEC/IEEE 42020 standard, an architecture view represents an "information item expressing the architecture from the perspective of specific stakeholders regarding specific aspects of the architecture entity and its environment", where the architecture entity refers to the "thing being considered, described, discussed or otherwise addressed during the architecting effort", i.e., in our context, the robot reference model. From the latter, assets and viewpoints are utilized to construct views using a process and based on a metamodel and notation.
- Assets refer to the total set of all model-related elements. Here, we distinguish between reference model assets, belonging to the reference model such as building blocks, patterns as well as individually defined architecture viewpoints. Domain assets, on the other hand side, define the assets, which are constructed using the

reference model and are utilized within a model configuration by the means of adoption, adaptation or extension of reference model elements.

- Parameters describe declarations for the possibilities how model elements can be instantiated in the reference model, given their specific types. For domain-specific modeling instead, they describe how these parameters are explicitly instantiated in a given model configuration.
- Building Blocks refer to the model elements themselves and their connections between them. TOGAF defines building blocks as potentially reusable components of business, IT or architectural capability, which can be combined with other building blocks to deliver architectures or solutions.
- Patterns are described as formations between different building blocks. As predefined sets of building blocks, they are intended to describe reusable configurations or formations of different building blocks. An architectural pattern to be a description of element and relation types with a set of constraints on how they may be used. Then a pattern can be considered a set of constraints on an architecture as well as on the element types and their patterns of interaction and constraints define a set or family of architectures that satisfy them.
- Repositories describe collections of all model-related artifacts belonging to a model, which define structure and format as well as location how this collection is stored. In the ISO/IEC/IEEE 42020 standard, a repository refers to a "place where work products and the associated information items are or can be stored for preservation and retrieval". In our context, these work products and information items relate to the reference model itself for the reference model repository or to domain-specific model configurations for the model configuration repository instead.

5.2 Signal View

For evaluation of reference model and domain-specific configurations instead, we describe the following concepts:

- Requirements represent the design goals of the system under design considered by the mathematics developed in Chap. 3. They are expressed with respect to a particular metric, which can be obtained from the system under designing such as particular parameters, structure and behaviors, thus they can exclusively or non-exclusively affect the configuration of structural model properties or behavioral model properties such as a pursuing a particular trajectory of states and actions. Based on model properties, requirements express a preference for achieving a particular metric and can be expressed either as objectives or as constraints.
- Objectives describe design goals in terms of the maximization or minimization of cost/utility functions based on a specific numeric metric.
- Constraints and specifications describe restrictions and operational limits on design goals. For evaluation, reference model assets are subject to general requirements,

which any sensible model design and the reference model itself should satisfy. Domain model assets on the other hand side are subject to the general constraints derived from the reference model itself as well as specific requirements, which only apply to a given domain or domain-specific scenario. For instance, this incorporates static requirements such as structural constraints describing valid arrangement of building blocks and model elements of a given type. Typically, these arrangements are quantified by global or existence constraints defining conjunction, disjunction as well as negation relations between given model elements. Then, according assertion rules about model element arrangement using these constraints are used for model verification. For defining objectives, we employ cost and contribution functions instead. Contribution and cost functions ascribe numerical values to model structure and behavior cost and contributions. Accordingly, goals are especially useful for comparing different design alternatives and choosing an alternative with the highest contribution or the lowest costs. According evaluations of model configuration alternatives are then established quantitatively based on the configuration space defined by the reference model. Evaluation is supported by simulation models as well as context/data generated by live system operation.

5.3 Physics View

The Field Robot is inserted into the real world as a physical entity, i.e. it is becoming part of the Random Dynamic System. This is a change of the model that the Designer uses, and by inserting some level of autonomy, kind of catalytically, there might become available another one. This is somehow describing the bottom-up way of finding the solution to the Random Dynamic System, now for the existence of an external force: the requirements as order for the Field Robot to obey. Looking from the top-view, the Field Robot is generating ROCs on a Probability of Performance versus Probability of Non-Performance diagram, whereby the hierarchy within the robots internal model (which is formulated similar to the Designer's Random Dynamical System's solution) adds more and more detailness (at the end of a consistent model to reality overlap: exactitude) to the resolution of this diagram. Remember that the systems solution are stepping hierarchically through a modelling with an increasing order of the Chaos Polynomials. As more details in the Designer's model lead to more details in the Robot's internal model, we need to make sure that this insertion in the insertion has a fixpoint, i.e. the level of details stop making a difference. On the other hand, if we start with a high level of detailness, we probably want to restrict our analysis to systems that do not suddenly break down if we abstract this highest level of detailness with some probabilistic description. With this, we are demanding that the fixpoint operation of giving either the robot or the designer access to more details is in its equilibrium.

In this equilibrium, as a restriction to the Design Space we explore, we can give the Designer the final say: The decision on the highest level of resolution. The number of decision levels and the number of options generated in the Random Dynamic System including a Field Robot, is then a metric for the difficulty of the given autonomous task: The hierarchical automatism calculating the solution for the Random Dynamic System calculates this metric automatically and independently from the way the Robot is actually implemented. Different bisimulations as autonomous approaches can then be evaluated as the robot's capacity.

How do these bisimulations look like? One theoretical option is to run the same Design Space Exploration as the Designer is using, but this has already been discussed as non viable. Another solution is to use an online version of the solution to the requirement-specification game that is the core of the Design Space Exploration the Designer is using, but it only works with the current situation the robot is in.

5.4 Online Requirement-Specification Game

The Design Space Exploration of the Designer delivers a set of equations describing the now goal-driven Random Dynamic System with the Field Robot as part of the system. Online Blind System Separation re-wires with the help of these equation its measurement in a way that the Field Robot in its own view becomes independent of the Dynamic System, which is no longer random in its structural parameters at its level of existence, because the independence is only valid for the current situation the robot is in and the (previously random coupling) parameters are estimated within the Blind System Separation. Once the Field Robot has this outer view on the task, it can start its own (limited to the current situation) Design Space Exploration with the iterated hierarchical solution for its planning and control, whereby the only explicit goal for the Field Robot is now to survive, i.e. to keep the separation status as long as possible.

What else has the Field Robot as equipment onboard? It can do BGoF in combination with FISST to estimate situational parameters. The situation is found via the best fit HMM, and then predicted within the best-fit HMM. The HMM algorithmic is modelled in the Design Space Exploration via INLA, whereby the INLA-driving noises have to be consistent with the noises calculated in the Cascaded Gaussian Functionals.

We postpone the description of another equipment onboard of the Field Robot at its conceptual level: Does the Field Robot decide to step down to further autonomous layers or will the task be already solved by the current level of programming (formulas can reach the level of DSE-global-detailness, but no new top-down-layer is automatically introduced by the Field Robot). For this decision-making we will introduce a POMDP in Sect. 5.5. In case no new level is decided for, an answer to the tasking layer for the Field Robot needs to be provided.

But from the discussion so far in this Section, we see that there are two different mathematical domains involved: the continuous domain, in which the Blind System Separation makes the Field Robots current abstraction level stay within uniformly distributed unknowns of some to be defined states and the Gaussian distributed influence from the activities one level lower than the current level; and the discrete domain, in which the states are selected (by active decision-making or by passively being driven within the Random Dynamic System). These two domains are linked by receiving noisy data about the state of an object, by controlling to pass a gate in some real or abstract sense. In this Section, we look at the two domains separately, as Online System Separation in Sect. 5.4.1 and as MAPE in Sect. 5.4.2. Then in the next section, we combine the two domains within a single algorithmic description which we call the Design Space Exploration-Machine (DSE-M), and we further investigate the implementation of this DSE-M algorithm.

5.4.1 Online System Separation

The dynamic behavior of the system from the Field Robots point of view is described by a non-linear differential equation. The system has two sources for the Field Robot: one that is playing it into the direction of the goal and (since the Field Robot is innocent of this, because it is inherently hidden in the equations submitted by the designer) has a uniform distribution, the second source is a Gaussian noise driving the Field Robot randomly and hiding its sensor measurements.

The explicit knowledge of the mixing coefficients is necessary if the current situation needs to be handled, because knowing the system parameters allows a short-term prediction of the systems behavior. Since we are talking about non-linear systems also chaos control becomes possible if the system equations are available. In the following, we outline a mathematical method which allows the Field Robot to separate itself explicitly from the system and thereby learning/estimating the system's parameters.

5.4.1.1 Mathematical Tool Set

In Sect. 2.2.3 we have used a moment based method to separate signals which have been mixed in a non-linear manner. We have used this technique at an abstract level to build up Cased Gaussian Functionals in Sect. 3.2. Now, in this Section, we are back in the domain of handling real signals in the robot.

Furthermore, procedures for the analysis and control of non-linear systems are needed, e.g. [6, 7].

5.4.1.2 Mathematical Description of the System Separation Task

We are assuming a Field Robot that has access to the mixed signals resulting from the sources it is linked to, and that the Field Robot can record these signals in its memory. Furthermore, we are assuming that the Field Robot's own behavior through its actuators is manipulating the system, i.e. in the process of Online System Separation we are feeding back the signals we are calculating inside the process into the system.

Online System Separation is calculating (from the time signals of the system it is linked to without a-priori knowledge on the coupling parameters) actuator signals as output, such that the dynamic system is manipulated in such a manner that the Field Robot becomes only an observer of the (manipulated) dynamic system, with its own internal processes separated from the dynamic systems.

Mathematically, this situation is described by:

1. **Subsystem** The complete system consists of $N \in \mathrm{N}$, $N \geq 2$ sub-systems. These sub-systems are (if running on their own) guided by the following equations:

$$\begin{aligned} x_1^{(t+1)} &= F_1\left(x_1^{(t)}, x_1^{(t-1)}, \ldots,\right) \\ &\vdots \\ x_i^{(t+1)} &= F_i\left(x_i^{(t)}, x_i^{(t-1)}, \ldots,\right) \\ &\vdots \\ x_N^{(t+1)} &= F_N\left(x_N^{(t)}, x_N^{(t-1)}, \ldots,\right), \end{aligned} \tag{5.1}$$

where $i \in 1, 2, \ldots, N$, the functions F_i are not further specified and unknown.

2. **Coupling of sub-systems** If the sub-systems are coupled, this relationship can be described by a number of coupling parameters such that (with coupling) the system is described by the following equations:

$$\begin{aligned} x_1^{(t+1)} &= F_1\left(x_1^{(t)}, x_1^{(t-1)}, \ldots,\right) \\ &+ W_1\left(x_1^{(t)}, \ldots, x_1^{(t-T_1)}, \ldots, x_N^{(t)}, \ldots, x_N^{(t-T_N)}\right) \\ &\vdots \\ x_N^{(t+1)} &= F_N\left(x_N^{(t)}, x_N^{(t-1)}, \ldots,\right) \\ &+ W_N\left(x_1^{(t)}, \ldots, x_1^{(t-T_1)}, \ldots, x_N^{(t)}, \ldots, x_N^{(t-T_N)}\right), \end{aligned} \tag{5.2}$$

where the functions W_i are a-priori unknown functions, having a number of parameters. Furthermore, the W_i shall not have terms which depend **only** on one system variable x_i.

From the theory of nonlinear dynamic systems (i.e. *Takens-Method* in [1] and from the nonlinear system theory (i.e. *Volterra series*) it is known that every dynamic system can be described by a set of parameters for the coupling functions W_i in Eq. (5.2) with an arbitrary exactitude.

3. **Online Blind System Separation** With the Online Blind System Separation, the a-priori unknown coupling parameters are estimated and with the estimated parameter values signals are calculated which de-couple the subsystems. With V_i

being the estimated coupling parameters, the effect of the decoupling is described by the following equations:

$$\begin{aligned}
x_1^{(t+1)} &= F_1\left(x_1^{(t)}, x_1^{(t-1)}, \ldots,\right) \\
&+ W_1\left(x_1^{(t)}, \ldots, x_1^{(t-T_1)}, \ldots, x_N^{(t)}, \ldots, x_N^{(t-T_N)}\right) \\
&- V_1\left(x_1^{(t)}, \ldots, x_1^{(t-T_1)}, \ldots, x_N^{(t)}, \ldots, x_N^{(t-T_N)}\right) \\
\vdots &\ \vdots\ \vdots \\
x_N^{(t+1)} &= F_N\left(x_N^{(t)}, x_N^{(t-1)}, \ldots,\right) \\
&+ W_N\left(x_1^{(t)}, \ldots, x_1^{(t-T_1)}, \ldots, x_N^{(t)}, \ldots, x_N^{(t-T_N)}\right) \\
&- V_N\left(x_1^{(t)}, \ldots, x_1^{(t-T_1)}, \ldots, x_N^{(t)}, \ldots, x_N^{(t-T_N)}\right).
\end{aligned} \tag{5.3}$$

If V_i are equal to W_i, then the sub-systems of Eq. (5.1) are generated.

4. **Estimation of de-coupling parameters** Short-term independence cost functions from received values can be used to estimate the best de-coupling parameters.

5.4.1.3 Main Idea

The main idea is to estimate de-coupling parameters, such that signals can be fed back into the system that separates the system into sub-systems.

Details on the De-coupling Parameter Estimation
It is assumed that the output signals of the sub-systems without coupling Eq. (5.1) are independent. The coupling generates a measurable dependence, e.g. for x_i, x_j from Eq. (5.2) we get:

$$\begin{aligned}
\langle x_i(t+1)x_j(t+1-\tau)\rangle &= \langle W_i(t)W_j(t+\tau)\rangle \\
&+ \langle F_i(t)W_j(t+\tau)\rangle \\
&+ \langle W_i(t)F_j(t+\tau)\rangle,
\end{aligned} \tag{5.4}$$

where $\langle\ldots\rangle$ means a sum over sufficiently many measured values.

The dependence can be rated via a cost function, e.g.

$$K = \sum_\tau \langle x_i(t)x_j(t+\tau)\rangle^2. \tag{5.5}$$

All cost functions that measure the dependence between $x_1(t)$ and $x_2(t)$ can be combined, in particular time-delayed cumulants (*the symbol for cumulants of order N is:* $\mathrm{cum}(x_1, \ldots, x_N)$).

Minimizing the cost function (e.g. numerically, or analytically as in the modelling of the Field Robot in the Design Phase) leads to an estimation of V_i, such that independent output signals $x_i(t)$ can be generated.

5.4.1.4 Implementation

As an example the nonlinear dynamic system:

- With initial values:

$$x_1[0] = 0.2;$$
$$x_2[0] = 0.3 \tag{5.6}$$

- and iteration steps:

$$\begin{aligned} x_1[i] &= \text{abs}(4.0x_1[i-1](1.0 - x_1[i-1])) \\ &\quad + 0.1x_1[i-1]x_2[i-1] - k_{1212}x_1[i-1]x_2[i-1]; \\ x_2[i] &= \text{abs}(4.0x_2[i-1](1.0 - x_2[i-1])) \\ &\quad + 0.2x_1[i-1]x_2[i-1] - k_{2212}x_1[i-1]x_2[i-1] \end{aligned} \tag{5.7}$$

  are defined to generate $x_1[i]$ und $x_2[i]$.
- Next, a strong nonlinearity is inserted to prevent chaotic behavior in the simulation.

```
if(x_1[i] > 2.0)
        x_1[i]=2.0;
else
{
        if(x_1[i] < -2.0)
                x_1[i]=-2.0;
}//else
if(x_2[i] > 2.0)
        x_2[i]=2.0;
else
{
        if(x_2[i] < -2.0)
                x_2[i]=-2.0;
}//else
```

  Note: This step is just preparation, not part of the online separation process.

The coefficients k_{1212} and k_{2212} need to be estimated, hence

$$k_{1212} = 0.1 \quad \text{und}$$
$$k_{2212} = 0.2 \tag{5.8}$$

Following Paragraph 5.4.1.2 we are using the cost function calculated from the output signals $x_1[i]$ and $x_2[i]$

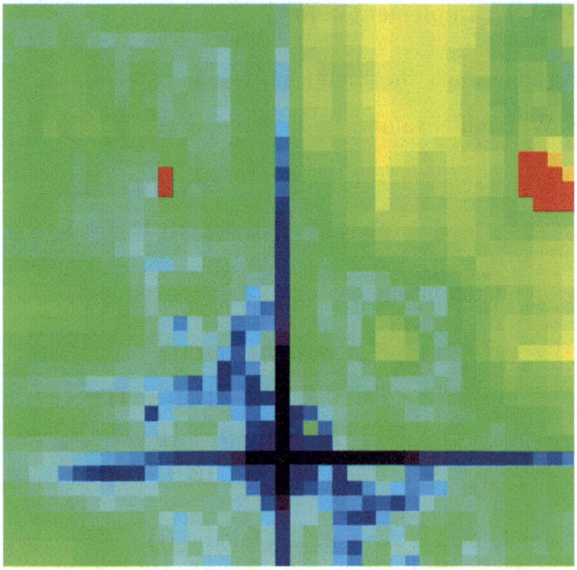

Fig. 5.1 Image of the cost function used in online separation of systems for the example in the Implementation Sect. 5.4.1.4

$$
\begin{aligned}
K(k_{1212}, k_{2212}) = {} & (\mathrm{cum}(x_1[i], x_2[i]))^2 \\
& + (\mathrm{cum}(x_1[i], x_1[i], x_2[i]))^2 \\
& + (\mathrm{cum}(x_1[i], x_2[i], x_2[i]))^2 \\
& + (\mathrm{cum}(x_1[i], x_1[i], x_1[i], x_2[i]))^2 \\
& + (\mathrm{cum}(x_1[i], x_1[i], x_2[i], x_2[i]))^2 \\
& + (\mathrm{cum}(x_1[i], x_2[i], x_2[i], x_2[i]))^2. \qquad (5.9)
\end{aligned}
$$

As seen in the Fig. 5.1, $(K(k_{1212}, k_{2212}))$ the global Minimum is at

$$
\begin{aligned}
k_{1212} &= 0.1 \quad \text{und} \\
k_{2212} &= 0.2 \qquad (5.10)
\end{aligned}
$$

5.4.1.5 Summary of Online System Separation

If the goal of the robot's operation is implicitly coded in the set of equations the Designer delivers to the Field Robot, following the planning and control solution brings the Field Robot automatically to the task solution. On the other hand, the online system separation is making the own survival of the robot become explicit.

If the Field Robot cuts itself at the level the cut is programmed for (i.e. the model of the Designer for the Field Robot takes the Online System Separation equation), it is necessary to get a consistent interface to global-DSE and local-DSE. The Robot needs to take the same noise assumptions than in the global DSE to estimate its state. We will come back to this once we have found the mathematical description for the noise assumptions.

Scaling of signals for the Online System Separation is necessary to get the Blind System Separation fixpoint algorithm work. Because the models for the signals to expect are available, such scaling can be done.

5.4.2 MAPE

This fits to the Measure Analyse Plan Execute (MAPE) process, which is a concept to implement autonomy in robots. The term "MAPE" refers to an acronym for four essential processes in robotics: Monitor, Analyze, Plan, and Execute. These processes collectively enable robots to adapt and respond effectively to their environments while ensuring optimal performance and decision-making capabilities. Here's a brief overview of each process: Monitor—The monitoring phase involves gathering data from various sensors and systems within the robot's environment. This information is crucial for understanding the current state, identifying potential challenges or obstacles, and recognizing opportunities to improve overall performance. Analyze—In this stage, robots process the collected data through algorithms and machine learning techniques to identify patterns, trends, and relationships between different variables. By analyzing the information, robots can better understand their environment and make informed decisions based on the insights gained from this analysis. Plan—Once the robot has analyzed its surroundings, it must develop a plan of action that considers multiple factors such as available resources, time constraints, and desired outcomes. This planning phase may involve generating a step-by-step sequence of actions or selecting an optimal path to achieve specific goals. Execute—Finally, the robot executes the planned course of action by implementing the selected plan in real-time while continuously monitoring its environment for changes or new information that might require adjustments to the initial strategy. This process ensures that robots can adapt and respond effectively to unforeseen challenges or opportunities as they arise during their operation. In summary, MAPE is a software framework used by robots to adapt and respond effectively to their environments through continuous monitoring, analysis, planning, and execution processes. By employing this approach, robots can optimize performance, make informed decisions, and ensure efficient navigation of complex scenarios in real-world applications.

It is important to note that the online separation solves the difficulty to run the entire Design Space Exploration on the Field Robot, but limiting its prepared capabilities to one situation at a time.

5.5 Design Space Exploration Machine (DSE-M)

The equations of the online separation are iterative, hence can again be incorporated in the overall set of equations, and their detailness in each processing step is automatically limited to the settings for the hierarchical level. Hence, it is possible to model the Field Robots internal calculation as part of the overall Random Dynamic System. Putting now together all elements the Designer has available, within the Architecture the Designer can generate all mathematical formulas describing the system with the robot in it, starting the optimization for the best setup for the robot by exploring the robot's performance in the analytically calculated ROCs.

5.5.1 Efficiency Optimisation

As written in the section on the online separation, the task goal has to be implicitly coded in the equations given to the robot. We can further limit the ways the robot can reach its goal by adding a constraint to only take the most efficient ways from the Design Space Exploration to the online operation. For this, we model the tree of options as a Poisson point process (PPP). A PPP is a type of stochastic process used to model events that occur randomly and independently within a given space or region. It is often used in fields such as statistics, mathematics, computer science, and engineering to analyze data sets with spatial or temporal components. In a PPP, the probability of an event occurring at any particular location is proportional to the number of events expected to occur over a fixed time interval, based on the average rate of occurrence across the entire space or region being studied. This means that the probability density function (PDF) for a PPP can be expressed as:

$$f(x;) = e^{(-\lambda x)}/x! \tag{5.11}$$

where x is the location where the event occurs, and λ is the rate parameter that determines the average number of events expected to occur per unit time.

Similar to an Expectation Maximization (EM) solution for such PPP given data, we can maximize the efficiency of the ways to reach the goal within the framework developed here with the trees of options representing the PPP process and the noisy endpoints of the tree representing overlayed Gaussian measurement uncertainties. The associated EM algorithm is developed in [4], and iterations can be inserted as analytic formulas in the mathematical equations developed so far.

5.5.2 POMDP as Conceptual BPMN

So far, we have generated a mathematical model for the Design Exploration process inside the Field Robot, where we can refer to the Decision-Making inside this Field Robot to be algorithmically done and described by a Decision Process on selection one level more of detailledness. The Designer has to model this ability, hence we need to investigate the best decision making plus the prediction of its performance. A Partially Observable Markov Decision Processes (POMDP) is suitable for this situation, remembering that we use it here at a conceptual level, i.e. within the concept generating layer of the Design Space Exploration-Machine.

Following [3], a POMDP is an advanced mathematical framework that helps artificial intelligence systems and other complex decision-making processes to handle uncertainty, partial observability, and dynamic environments by formulating optimal strategies based on available information and potential outcomes.

A partially observable Markov decision process (POMDP) is a combination of an regular Markov Decision Process to model system dynamics with a hidden Markov model that connects unobservable system states probabilistically to observations.

The agent can perform actions which affect the system (i.e., may cause the system state to change) with the goal to maximize the expected future rewards that depend on the sequence of system state and the agent's actions in the future. The goal is to find the optimal policy that guides the agent's actions. Different to MDPs, for POMDPs, the agent cannot directly observe the complete system state, but the agent makes observations that depend on the state. The agent uses these observations to form a belief about in what state the system currently is. This belief is called a belief state and is expressed as a probability distribution over all possible states. The solution of the POMDP is a policy prescribing which action to take in each belief state. Note that belief states are continuous resulting in an infinite state set which makes POMDPs much harder to solve compared to MDPs.

A discrete-time POMDP can formally be described as a 7-tuple $P = (S, A, T, R, \Omega, O, \gamma)$, where $S = s_1, s_2, \ldots, s_n$ is a set of partially observable states, $A = a_1, a_2, \ldots, a_m$ is a set of actions, T is a set of conditional transition probabilities $T(s'|s, a)$ for the state transition $s \rightarrow s'$ conditioned on the taken action.

$R : S \times A \rightarrow R$ is the reward function, $O = o_1, o_2, \ldots, o_k$ is a set of observations, O is a set of observation probabilities $O(o|s, a)$ conditioned on the reached state and the taken action, and $\gamma \in [0, 1]$ is the discount factor.

At each time period, the environment is in some unknown state $s \in S$. The agent chooses an action $a \in A$, which causes the environment to transition to state $s' \in S$ with probability $T(s'|s, a)$. At the same time, the agent receives an observation $o \in \Omega$ which depends on the new state of the environment with probability $O(o|s', a)$. Finally, the agent receives a reward $R(s', a)$. Then the process repeats. The goal is for the agent to choose actions that maximize the expected sum of discounted future rewards, i.e., she chooses the actions at each time t that $max\, E[\sum_{t=0}^{\infty} \gamma_t R(s_t', a_t)]$.

For a finite time horizon, only the expectation over the sum up to the time horizon is used.

The automatic relationship between the description of the POMDP and the predicted control/performance/goal achievement variables has to be coded by the Designer, but this is a natural relationship in the sense that the decision to wait for more details to explored or calculated is already directly related to the risk to miss a goal by this decision. Since the Designer has a holistic view, it is possible to assign values according to the balanced game which Requirements versus Specifications play.

Take-Away for Design Space Exploration
In the previous chapter, we have identified that all details of the Design Space can be stored in the architectural framework. In this Chapter we are able to model the Field Robot becoming a part of this Design Space for the Designer, but for the robot itself its behavior can be generated in a computational viable manner, in a predictable manner wired to the details stored in the architectural framework, and with the online system separation technique it is possible to switch from explicit goal description in the requirements to explicit goal description in terms of the robots independence of the outer system, where independence here means that neither the preset goals nor the robot itself can get harmed by the autonomous operation designed.

References

1. Takens, F.: Detecting strange attractors in turbulence. In: Rand, D.A., Young, L.S. (eds.) Dynamical Systems and Turbulence, Warwick. Lecture Notes in Mathematics, vol. 898, pp. 366–381. Springer, New York (1980)
2. Ehlers, F.: Acoustic Modelling Knowledge for Design and Specification of Situation-Adaptive Distributed Networked Systems, UACE 2019
3. Partially Observable Markov Decision Processes. https://cran.r-project.org/web/packages/pomdp/vignettes/pomdp.html. Cited 2024/07/12
4. Streit, R.L.: Poisson Point Processes. Springer (2010)
5. Ascher, D., Ehlers, F.: Seamless integration of ATR services into situation-adaptive distributed systems via reference modeling. Proc. Mtgs. Acoust. **44** (1), 070020. https://doi.org/10.1121/2.0001487. Cited 2021/06/21
6. Pyragas, K.: Continuous control of chaos by self-controlling feedback. Phys. Lett. A **170**, 421–428 (1992)
7. Ott, E., Grebogi, C., Yorke, J.A.: Controlling Chaos. Phys. Rev. Lett. 64, 1196–1199 (1990)

Chapter 6
Robot-Robot Coordination

Abstract In the previous chapter we elaborated on the inner and outer views of the Designer and the Field Robot. In this Chapter, we discuss consequences from this in the framework of multi-robot coordination. Also, in the previous chapter we identified the connectivity between the POMDP and the conceptual decision making on the level of detailness the Field Robot can choose in the online operation. However, as we will see from the discussions in this Chapter, the POMDP formulation is not sufficiently generic, in terms of the formulation of the process of finding solutions for it. Therefore, we identify the Potts model as a generic replacement, which finally allows us to formulate a completely holistic model from which we will derive autonomy metrics in the next chapter.

Top-Down View
Almost there, at the generic formulation of the Design Space Exploration in Robotics. But we can already imagine that a POMDP with an iterative solution finding with specific problems with local minima and computational load for large number of options and state spaces might be hindering a viable implementation. The discussion of multi-robot coordination will point to a replacement algorithm.

From the first part of the Introduction we record, that in Chap. 6 we will put our focus on the overlapping area of the two approaches (bottom-up, top-down) for the specific application of Multi-Robot Coordination. There, if we see a mismatch between the results of the two approaches, decisions have to be made with risks/costs following the performance metric we developed in Chap. 5. Influence Diagrams are a graphical way to describe this decision making. The values for the Influence Diagram are generated by the so-called Validation and Verification (V&V) process. Whilst Validation and Verification are representing the top-down and bottom-up directions, respectively, in a situation where bottom-up and top-down approaches match without

F. Ehlers, *Design Space Exploration in Robotics*, Studies in Systems, Decision and Control 229, https://doi.org/10.1007/978-3-031-78740-9_6

inconsistencies, it cannot be forgotten that the system has to be implemented, i.e. becomes part of a real system with real noise, real hardware failures, real timing, etc. The Certification process, as an outer view on the designed system, continues the system analysis with the assumption that inconsistencies occurring inside the Field Robot (which might originally be originated also by non-resolved, but in the light of the given task acceptable inconsistencies between bottom-up and top-down approaches) are belonging to the discrepancy between measurement interpretation and reality, i.e. by the fact that within a Field Robot only an estimate of the a-priori probability can be extracted from data. For a technically engineered system then, we add a "final checking device", which is equipped with model information and data, to perform Certification as a final check, e.g. via formal models regarding externally and prior formulated certification rules. Coming to the final acceptance test, we need to apply a no-trust methodology and generate statistically significant data to make a final decision on whether the entire system is fulfilling its requirements, whereby no-trust means that we can only trust the measurements in the real world, not models nor simulations.

6.1 Influence Diagram

Following [2], Influence Diagrams are utilized for the task of decision making using decision-theoretic principles within the graphical modeling framework. Whereas a purely probabilistic graphical model provides a factorized description of the probability distribution over possible state of world, an influence diagram provides such a factorized representation for the agent's action and utility function as well. The influence diagram cleary encodes the breakdown of these three components of the decision-making situation in variables, as well as the interactions between these variables. These interactions are both probabilistic, where one variable affects the distribution of another, and informational, where observing a variable allows an agent to actively change his action (or his decision rule). The complexity is due to the interactions the different operations involved: products for defining the probability distribution; summation for aggregating utility variables; and maximization for determining the agent's optimal actions. Value of information as a concept plays a key role in many practical applications and is particularly easy to capture in the influence diagram framework.

The idea is to prepare different strategies for setup Design Space Exploration-Machines, and use the influence diagram to display and calculate the effects. The strategies in Design Space Exploration therefore point to different references, and they give the Designer a clue on how to continue in the Design Space Exploration. Does it pay off to limit Validation and Verification to a specific detailness in the robot's DSE-M, etc.? are questions to put into this abstract level of decision making.

Within the framework of multi-robot coordination these questions can be related to: How much modelling needs one robot's behavior in the other robot's behavior? Clearly, iterating dependencies will lead to very complicated calculations, probably with limited effect on reaching the goal, or rather decreasing the performance due to overwhelming processing needs. Hence, we need to look for a formulation where the overall performance regarding the Gaussian ROCs from the tree of options has to be linked to the decision making in the influence diagrams.

6.2 Validation and Verification

The PMBOK guide, a standard adopted by the Institute of Electrical and Electronics Engineers (IEEE), defines them as follows in its 4th edition:

"Validation. The assurance that a product, service, or system meets the needs of the customer and other identified stakeholders. It often involves acceptance and suitability with external customers. Contrast with verification." "Verification. The evaluation of whether or not a product, service, or system complies with a regulation, requirement, specification, or imposed condition. It is often an internal process. Contrast with validation."

In multi-robot coordination the usage of verified and validated behavior of each robot is crucial when building the Design Space Exploration-Machines for them. Since robots will have the ability to react/adapt on online generated unplanned behavior in an autonomous manner, the overall outcome of the system will degrade quickly.

6.3 Certification

Certification is what the system is showing to the outside: Does it damage something? Does it misbehave to collaborating external interface in the sense of externally given constraints? A certification comes in later, after formal verification, since it is the interaction at realisation-time with the environment. V&V is at design-time. Externals to the design-process perform the certification, and certification requirements can also change after the design of a product has been already finalized.

Formal verification is not synonymous with certification but rather serves as a crucial component in the process of certifying complex software or hardware systems. While formal verification focuses on applying mathematical techniques, logical proofs, and rigorous analysis methods to ensure that a system meets its specified requirements and functions without errors or vulnerabilities, certification refers to the official recognition or approval granted by an authoritative body or organization after verifying compliance with specific standards, regulations, or guidelines. In summary, formal verification is one of the key steps in obtaining certification for complex software/hardware systems but should not be considered equivalent to it.

Formal verification is the process of applying mathematical techniques and logical proofs to establish that a given complex software or hardware system meets its specified requirements, adheres to design constraints, and functions as intended without introducing any errors or vulnerabilities. This rigorous approach helps ensure high-quality products with minimal risks associated with security breaches, performance degradation, or other undesirable outcomes. There are several formal methods that can be employed for certifying complex software/hardware systems:

- Formal Models—These involve creating abstract representations of the system in question using mathematical languages such as Z, B, or VDM to model its structure and behavior. This allows for a more systematic analysis of the system's properties and enables verification through theorem proving techniques.
- Theorem Proving—Utilizing automated tools like HOL Light, PVS, or Isabelle, theorem provers can be employed to prove that specific properties hold true within the context of the formal model. This process helps identify any inconsistencies or errors in the system's design and ensures compliance with its intended functionality.
- Model Checking—Techniques like Binary Decision Diagrams (BDD) and Symbolic Execution can be employed to explore all possible states, transitions, and paths within a given model of the system. By exhaustively examining these scenarios, model checkers can identify any discrepancies between the expected behavior and actual outcomes.
- Runtime Verification—This approach involves monitoring the system's execution in real-time using techniques like runtime assertion checking or invariant tracking to ensure that it adheres to its specified requirements throughout its operation. By detecting deviations from the desired behavior, runtime verification can help identify and rectify errors as they occur.

In summary, employing formal methods and other logical checks is essential for certifying complex software/hardware systems by ensuring their correctness, reliability, and security in various operational scenarios. Through a combination of mathematical modeling, theorem proving, model checking, and runtime verification techniques, organizations can build trustworthy products that meet the highest standards of quality and performance.

In the context of insurance, certification plays a vital role in ensuring that companies adhere to regulatory requirements and maintain high standards of service quality. Certification typically involves obtaining licenses or accreditation from relevant authorities, which demonstrate compliance with industry-specific rules and guidelines. This process helps build trust between insurers and their clients by providing assurance that the insurance provider is competent, reliable, and capable of fulfilling its contractual obligations. Certification also serves as a means for regulators to monitor and supervise the activities of insurance companies, ensuring they operate ethically and responsibly while safeguarding policyholders' interests. Additionally, certification can be beneficial in promoting transparency and accountability within the industry by providing a framework for evaluating insurers based on their performance, financial stability, risk management practices, and customer satisfaction levels. In summary, certification is an essential component of the insurance landscape as it ensures that companies adhere to regulatory standards, maintain high-quality service, and provide clients with confidence in their chosen provider's ability to meet contractual obligations.

In the framework of the DSE-M, we can model the capability of the Field Robot regarding its parameter estimation (BGoF and FISST are analytically implemented) and decision making (HMM via INLA is analytically implemented). Hence, all values needed to perform Certification can be extracted from these calculations. Where computationally limits occur, the effects of further approximations can be predicted in the framework of the influence diagrams.

6.4 Acceptance Test

Acceptance test for autonomous systems should indeed verify that the system meets all its specified requirements and adheres to design constraints. This involves employing a combination of formal methods such as formal models, theorem proving, model checking, and runtime verification techniques to ensure:

- Compliance with functional specifications—The system must perform its intended tasks accurately and efficiently without any errors or discrepancies in the execution process.
- Adherence to non-functional requirements—These include aspects such as performance, reliability, scalability, security, maintainability, and usability of the autonomous system. Ensuring that these properties are satisfied is crucial for a successful deployment.
- Compliance with safety constraints—Autonomous systems often operate in environments where human lives or critical infrastructure may be at stake. It's essential to verify that the system can safely handle edge cases, unexpected inputs, and potential failures without causing harm or damage.
- Conformance to regulatory standards—Depending on the industry and application domain, autonomous systems might need to comply with specific regulations or guidelines. Ensuring adherence to these requirements is crucial for obtaining necessary certifications and licenses.
- User satisfaction—Lastly, acceptance tests should also consider user feedback and preferences in terms of usability, intuitiveness, and overall experience interacting with the autonomous system. This aspect can be addressed through surveys, focus groups, or other forms of user-centered evaluation methods.

The observation reading through the demands of an acceptance test emphasize that we cannot afford having machine learning techniques involved in our DSE-M, which is the POMDP. We cannot guarantee its performance in all cases without overwhelming computation need. Hence, we have to look for its generic replacement.

6.5 Generic Replacement of POMDP

In order to find the replacement for the POMDP, we refer to the El Foral bar problem as described in [3]. Let's discuss the problem first, and then transfer the findings into the Design Space Exploration domain.

- Problem: The El Farol bar problem is a "minority game". If too many people attend the bar, the patrons will not enjoy the evening. The problem for each potential visitor is to find out in advance whether the bar will be crowded below 50% or not. Therefore each of the N possible visitors i must decide whether to go ($\sigma_i = 1$) or stay home ($\sigma_i = -1$). Only those agent who are in the minority, i.e. who went, when the bar was less than half full, or stayed home when the bar war overcrowded, win. Decisions are made independently, i.e. the agents cannot communicate with each other directly. The only information available to them is the decision of the minority in the last L time-steps which determines the reward which they receive, after having made their decision. Similar problems arise in the context of an economy where traders try to find the least crowded, but still promising, stock on the market. In [3], Schuster models each agent by an adaptive probabilistic Boolean function with L inputs. The assumption is that there is an odd number of N players, each of which receives a binary input string $\mu = (x_1, ..., x_L)$, where $x_\alpha = +/-1$. Each player i reacts to this input with a binary output σ_i which becomes $\sigma_i = +/-1$ with the probability

$$P_\mu = 1/2(1 + \sigma_i \tanh(w^i_\mu)) \tag{6.1}$$

  The 2^L adaptive weights w^i_μ determine which random Boolean models player i. If the player i who is stimulated at time t with μ reacts with an output σ and receives a reward $r(\sigma_i, \mu)$ his weight w^i_μ changes according to the learning rule

$$\delta w^i_\mu = \eta r(\sigma_i, \mu)(\sigma_i - <\sigma_i>) \tag{6.2}$$

  where η determines the learning rate. This means that the weights w^i_μ increases if he received a positive reward and it stops changing if σ_i equals the average value $<\sigma_i> = \tanh(w^i_\mu)$.
- Idea for solution: The idea is to introduce a reward function which measures the deviation from the minority and show that the cot function of the minority problem is the energy function of an all-to-all coupled antiferromagnet.
- Solution: At time t the difference between the number of people in the bar and those who were staying home is

$$N^t_+ - N^t_- = \sum_{i=1}^{N} \sigma^t_i \tag{6.3}$$

In the minority game each player receives at time t the same input $\boldsymbol{\mu}$ and should adjust his weights in such a way that the sign of his output σ_i is opposite to the sign of majority $\sum_{i=1}^{N} \sigma_i$. In the language of the original El Farol bar problem this means, he should have gone to the bar ($\sigma_i = 1$) or stood home ($\sigma_i = -1$) only if it was less (more) than half full, i.e. if $sign(\sum_{i=1}^{N} \sigma_i) = -1$ or $(+1)$. Accordingly his reward r becomes, for a specific input $\boldsymbol{\mu}$, which specifies σ_i,

$$r(\sigma_i, \boldsymbol{\mu}) = -\sigma_i(\sum_{j=1}^{N} \sigma_j). \tag{6.4}$$

The goal of the game is to maximize the overall averaged reward

$$\frac{1}{2^L N} \sum_{i,\mu} \langle r(\sigma_i, \boldsymbol{\mu}) \rangle \tag{6.5}$$

or to minimize the cost function

$$C = \frac{1}{2^L N} \sum_{i,\mu} \langle r(\sigma_i, \boldsymbol{\mu}) \rangle = \frac{1}{2^L} \sum_{\mu} \langle \frac{1}{N} (\sum_{j=1}^{N} \sigma_j)^2 \rangle. \tag{6.6}$$

The term $\frac{1}{N}(\sum_{j=1}^{N} \sigma_j)^2 = \frac{1}{N}(\sum_{i,j=1}^{N} \sigma_j \sigma_i)$ describes the energy of an all-to-all antiferromagnetically coupled Ising model with coupling $J = 1$. This means the task of the agents is to adjust their Booleans in such a way that the averaged energy of this aniferromagnet will be minimized. By performing the average $\langle \ldots \rangle$ with $\Pi_{i,\mu} P_\mu(\sigma_i)$ the cost function becomes

$$C = 1 - \frac{1}{2^L N} \sum_{\mu} \left(\left(\sum_{i} (\tanh(w_\mu^i) \right)^2 - \left(\sum_{i} \tanh(w_\mu^i))^2 \right) \right). \tag{6.7}$$

With the assumption that in the learning equation the change of weights occurs slowly as compared to the spin flips, working with the average becomes possible as well:

$$\Delta w_\mu^i = \eta \langle r(\sigma_i, \boldsymbol{\mu})(\sigma_i - \langle \sigma_i \rangle) \rangle \tag{6.8}$$

$$= -\eta (1 - (\tanh(w_\mu^i))^2) \sum_{j \neq i} (\tanh(w_\mu^j)) \tag{6.9}$$

The right hand side is proportional to $\partial C / \partial w_\mu^j$, i.e. the function C is a Liapunov function for the average learning dynamics of the weights and its fixed point are the minima of C.

The antiferromagnetically all-to-all coupled Ising model with N spins has $M = N!/(((N + 1)/2!((N - 1)/2)!)$ different groundstates. The dynamics for different inputs $\boldsymbol{\mu}$ are uncoupled and the equations of motion could at modest detect 2^L different minima. For the case $L = 0$ where each player has only one weight w^i, the groundstates correspond to strategies where at most $N/2 + 1$ players will attend the bar at the same time, i.e. the cost function reduces to $C_min = 1/N$. But the strategies are fixed by $\{w^i\}$, i.e. always the same people will visit the bar. This changes already for $L = 1$. In this case two goundstates will be found by the dynamical equations. Random initial conditions $\{w^i(t = 0)\}$ generate in the long time limit different solutions $\{w^i_{\mu=1}\}$ and $\{w^i_{\mu=-1}\}$ and different inputs $\mu = +/-1$ will generate jumps of $\boldsymbol{\sigma}(t) = (\sigma_1, \ldots, \sigma_N)$ between these two different groundstates. This means that two different optimal populations will visit the bar randomly in time. Generally $2^L \leq M$ minima will be visited, triggered by different input patterns $\boldsymbol{\mu}$.

- Summary El Farol bar problem: By modelling different agents by different random adaptive Booleans Schuster has found a solution to the El Farol bar problem which reduces the fluctuations around the optimal number of visitors to their minimal value $C = 1/N$. His approach maps the solutions to the minority problem to different energy minima of an all-to-all coupled antiferromagnet and different inputs to the Booleans generate jumps between these minima which ensure that different optimal populations will visit the bar.

The choice of the agent in the calculation of the El Farol bar problem is similar to the POMDP choice we need to generalize: Staying in the same level of detailness or go one level deeper. With the formulation of the reward function for the all-to-all coupled antiferromagnet (which needs an extension to the reward function of an all-to-all coupled Potts model for indicating in which option of the deeper level the agent is going to) we have an additional potential function which we can overlay over the holistic DSE cost function of the requirement-specification game. The Ising Model and Potts Model are both statistical mechanical models used to study phase transitions and critical phenomena in disordered systems, specifically those involving spins or variables at each site. They share similarities but also have key differences: In the Ising Model, each site can take only one of two possible states (spin = $+1$ or -1), while the Potts Model allows for q different spin states. This means that the Potts Model is more general and can be used to study a wider range of systems with multiple types of interactions between spins. There are theoretical limits to produce stable/balanced Potts models, and these limits can be added to the optimization potential as constraints, hence the overall approach of setting up a complete analytical approach is not harmed by the switch from POMDPs to Potts models. Instead, because we have limited our Tree of Options to stop at Gaussian noise sources, the level-wise construction of stable Potts models in an analytic manner is possible.

Furthermore, introducing the Potts model's potential means, we restrict within the Design Space Exploration the online-operating Field Robot to stay in the area of its online operation (instead of porting the capability of performing an overall holistic DSE onboard the Field Robot which is not feasible in term of computational power/energy consumption). All mathematical formulations we are using rely now on the same mathematical foundation: a Gaussian noise driven hierarchical tree of decision making at increasing level of detailness, until the details do no longer count for reaching the given operational goal. We use the same formulations inside the robot and outside in the Design phase, in the models the Robot generates by itself and in the models the Designer generates for the model. All models can be solved analytically (with the El Farol bar problem solution the Field Robot would perform an agent-transport problem, in the Designer's model for the Field Robot the analytic behavior of such transport model is analysed) and come up with calculations respecting all details of the system and environmental models available.

Take-Away for Design Space Exploration

The Potts model speaks, so to say, the same language as INLA and HMM in the inner-view and as PPP and POMDP in the outer-view of the Field Robot. That is, it is possible to write down one global optimization cost function. This is a result for which we we have identified as a necessity to transfer this cost function to a Design decision-making. Therefore the Potts model serves as a glue within the seamless completely analytical and generic model we are looking for. In the next chapter, we use this model to generate the autonomy metric.

References

1. Ehlers, F.: Cooperative vehicle target tracking. In: Curtin, T., Dhanak, M.R., Xiros, N.I. (eds.) Springer Handbook of Ocean Engineering, Part B: Autonomous Ocean Vehicles, Subsystems and Control, Chap. 22. Springer International Publishing (2016)
2. Koller, D., Friedman, N.: Probabilistic Graphical Models. MIT Press (2010)
3. Schuster, H.G.: Complex Adaptive Systems. Scator Verlag (2001)

Chapter 7
Human-Robot Coordination

Abstract After being able to automatically generate a generic solution for the Design Space Exploration in Robotics in the previous chapters, the remaining task is to separate metrics for the autonomous approach from the metrics for the difficulty of a given task, where separation is meant to look for a formulation in which the probability of fulfilling a task by the Field Robot has two independent components, difficulity of the task and capacity of its autonomy.

Top-Down View

The primary distinction between robot-to-robot coordination (R2R) and human-robot coordination (HRC) lies in the nature of interaction, communication, and decision-making processes involved.

- Interaction: In R2R scenarios, robots communicate with each other using predefined protocols or algorithms to coordinate their actions, movements, and tasks. This can involve sharing information about positions, goals, resources, or constraints in the environment. On the other hand, HRC involves a human-robot interface where humans provide input, instructions, or feedback while robots execute tasks accordingly.
- Communication: In R2R coordination, communication typically occurs through standardized protocols or shared memory structures that enable efficient information exchange between robotic agents. Conversely, in HRC scenarios, communication can be more complex and multifaceted, encompassing both verbal and non-verbal cues, gestures, and signals to facilitate a seamless interaction experience.
- Decision-making: In R2R coordination, decision-making is often based on predefined rules or algorithms that determine the optimal course of action in response to environmental changes or other robots' actions. In contrast, HRC requires adaptive and responsive decision-making processes that can accommodate human input, preferences, and behavioral variations while ensuring safe and efficient coordination between humans and robots.

F. Ehlers, *Design Space Exploration in Robotics*, Studies in Systems, Decision and Control 229, https://doi.org/10.1007/978-3-031-78740-9_7

- Complexity: R2R scenarios tend to be more straightforward as they involve a closed system of interactions among robotic agents. In contrast, HRC situations are often more intricate due to the need for effective communication, understanding human intentions, and ensuring safety in shared environments.

In summary, robot-to-robot coordination focuses on inter-robot communication and decision-making processes within a closed system of interactions, while human-robot coordination involves complex interaction, communication, and adaptive decision-making to ensure safe and efficient collaboration between humans and robots in shared environments.

Finalizing the story-telling of the first part of the Introduction, contrasting in this Chapter the view on ourselves as humans equipped with a free will (or as non-technically engineered objects) and the view on the Field Robots with an only as-good-as-possible level of top-down/bottom-up consistency, we as humans seem to be standing outside the processes listed in Chap. 6, leaving the work to be done there in principle to computers and their artificial intelligence. But is this purely technical view physically and ethically viable? As computations become expensive for complex systems, there will be a maximally affordable degree of resolution with an associated tolerance of system applications without success. We have identified that we aim for systems that degrade gently with restriction to lower level of detailedness. We would call a system robust, if this gentle degradation holds. Instead if we see that little details start counting for the system evaluation, the system evaluation is loosing robustness.

7.1 Autonomy Metric

What have we described so far? A Field Robot that is equipped with a method that is capable to order incoming data by rotating and scretching model input sources consisting of two uniformly distributed independent noises in such a manner that we have a scale-invariance regarding the detailness used for a task performance measure related to the incoming data, whereby with more detailness the two noise sources' effects diverge in uniform distribution shape and a Gauss distribution shape, stopping at a certain level of detailness at which the requirements for the robot's task are fulfilled. The Field Robot is locally capable to decide for the most efficient way to proceed. Hence, the outside picture on the Field Robot is that the Field Robot might encounter various situation in which the Field Robot finds itself at a specific ROC-partition. Fulfilling a continuous sequence of incoming autonomous tasks, the Field Robot needs to combine these ROC-partitions to an overall ROC. Calculating this overall-ROC would be possible from the Designer's point of view by integrating over all possible tracks/paths of situations that might occur whereby fusing all sensor data on the system. This kind of bottom-up ROC generation is practically infeasible due to the high number of possible combinations of states. Luckily, the Designer has already a global DSE formulation, in which the Field Robots capabilities are contained and properly modelled. (Due to the incorporation of automated prior probability density estimation (BGoF) and the automated decision-making on the sufficient level of details to work with by the Field Robot itself, the learning-from-data capacity of a Field Robot is incorporated in this mathematically sound description of the Field Robot.) And even more, this global DSE formulation has a built-in scale-invariance, because the level of detailness we started with has been successively increased. Hence, we can extrapolate from the stopping level to the overall performance (which has already been a constraint to the equation-generation process of the Field Robot which is a step-wise top-down structural development process incorporating all design options and related parameters). Hence, the connectivity between top-down and bottom-up design processes is inherently (as much as possible) granted. To describe the Field Robot's performance the sequence of ROCs (starting from the most fine-grained level) the critical area of parameters is found by overlaying the ROCs of all layers, that is scale-invariance for all coarse-graining levels. Departing from the critical area will result in an exponential increase/decrease of performance of the system. Following [1], we have now a new tool at hand to describe the Field Robot: We have transformed the Design Space Exploration task into a mathematical domain which is capable to inspect and describe critical phenomena. In other words, we are able to describe the Field Robot in its various design options like a macroscopic system. Think about other macroscopic variables like specific heat, magnetic susceptibility or correlation length and their dependence on the temperature approaching the critical temperature. We can know describe the Field Robot's performance based on critical parameters in its behavior description: for example, a critical acoustic noise energy related to the wind turbulances related to the Field Robot's velocity, or a critical noise level from random non-flat terrains, or a critical error level in robotic communication networks. The long story of collecting details of specific sensors and actuators, their noisy measurement environment and interaction with the real world, which we have told in the previous chapters is now transferred into generally applicable descriptive variables for the Field Robot. These variables have an approximative analytic description (approximation as a critical system with exponential parameters) which is mathematically linked to the detailed setting due to the restriction of the separation models to Gaussian and Uniform distributions for which so-called Renormalisation techniques can be applied to derive the approximative formulas (Fig. 7.1).

We have prepared the following discussion in this Section by mentioning in previous chapters, that we restrict our Design Space Exploration to systems in which the level of detailness can be renormalized by coarse-graining and while doing this, the system performance shall not suddenly degrade. A robust, smoothly balanced system seems to be the most reliably Design Space Exploration output which the DSE-M can offer. But the demand for robustness is not the only design force: The

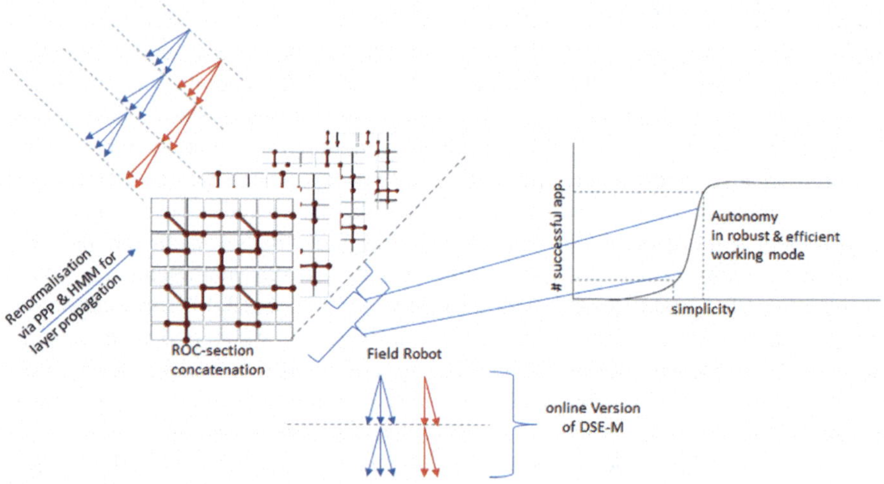

Fig. 7.1 Overview of the approach: A top-down hierarchy (upper left corner) sees at different levels of resolution scales design options and their estimated performance (center). The Field Robot (bottom center) separates a unique design from the options and performs at the right level of complexity. Autonomy Metrics are related to the shape of the curve on the right showing the relationship between number of successful applications of the autonomy and the simplicity of the actual scenario in which the autonomy had to perform the given task

Field Robot shall be as much as possible efficient. Hence, The designer will drive the Robots design always towards its critical parameters, i.e., choosing a detailedness that is just sufficient, performing data analysis that has just sufficient significance. The same capacity is prepared within the Field Robot by the programming techniques of maximizing efficiency and minimizing the need for data.

As a market is an evolutionary system for finding a global optimum in design, this balance between robustness and efficiency can be linked to the balance between requirements and specification (leading to the Global Optimization of Design Space Exploration): The market is driving to the a most robust requirement formulation and most efficient specification. The Enterprise Architecture naturally enters the scene as a management tools to guide the overall design towards this non-cooperative game maximization. Separating requirements and specification has been a driving idea for the Design Space Exploration approach in this Book. The fusion in the manner organized by an Enterprise Architecture justifies the incorporation of this abstract tool because the fusion result become visible and usable in the same tool that is used for describing all input knowledge and performing the model-based code generation. In the formulation of the Blind Separation cost function nothing changes, but there are previously fixed values for the requirement and specification that are now being enabled for the optimization process. By performing this global Design Space Optimization process we are encountering tasks (i.e. requirement/specification settings) at a different level of difficulty. By observing that easier tasks can be solved in coarser grains of detailness, whilst more difficult tasks need to handle more levels of details, we can (for example) use the number of levels of details needed in the Tree of Options to reach the operational goal as measurement of how difficult the task is. More generally, as we are working in the environment of coarse-graining renormalisation for Dynamical Systems we can link to a small number of critical exponents describing the behavior of the overall system, or in other words, with the rigorous formulation of the Design Space Exploration task as a Gaussian noise driven hierarchical system we have a renormalisation methods at hand to describe the difficulty of the task like Deep Neural Network informational capacity is estimated.

Solving the most difficult design task automatically describes a critical surface: If you try more, the performance will drop. Hence, now, at these critical points of the Dynamic System (i.e. the Field Robot in its local environment) we can further examine the way the overall performance is reached: Time-scale, length-scale, robustness to noise, consumption of energy. All tools from the renormalisation theory are available.

Furthermore, if we do this marginalising over the tasks (or categories of tasks), we can generate measures only depending on the setup of the autonomy, i.e. we can develop metrics for the autonomy. On the other hand, when the Field Robot is implemented with its autonomous features by different setting of the DSE-M, also these different settings and their output can be compared by calculating the output of the associated influence diagrams. Hence, both, task and autonomy, system-wide performance formulations via renormalisation techniques can be used to predict the probability of finding a solution for the task analytically and with independently set noise sources for the automous robot. Hence, autonomy metrics like the susceptibility to noise or predictability of behavior over longer usage times can be calculated.

Finally contrasting the bottom-up approach and the top-down approach, the bottom-up approach would be to start at the lowest level to disconnect the Field Robot. Once this is done, the designer sees this new robot and needs to program it within a new DSE: i.e. everything up to this level has to be re-designed. This is different from the top-down approach, where only one level is receiving a cut, and then, the Field Robot internally takes on to manipulate further with its internal DSE: The designer has no need to re-check this, because due to the system independence the robot is not harming the previous DSE.

In the real-world we cannot freely choose the design of each level, but we have to take as a tool/equipment, what is best and available, or already specified to be incorporated. Hence, instead of the free choice in the lowest/stopping level we just assign the best available asset, and adding its capabilities as a limiting constraint to the global DSE equations. The way we make this choice is the strategy: It is just like strategically decide on how to write a numerical optimization algorithm: search intervals, stepsize, and it might contain an iteration in the sense that the first steps have been too large. The core idea for dealing with the real-world constraints is that the robustness restriction of a balanced detailness is allowing this bottom-up parameter search: without this constraint a complete overlap would be needed in the limits of performance constraints, which would probably result in a never ending and highly dynamic bottom-up/top-down wave process. By introducing a level of allowed imperfectionism, i.e. an area for the ROC including errors, we are damping this wave process towards a robust solution. And in fact, technically designed and implemented systems never reach 100% error-free performance. The distance of the chosen parameters from the critical area found in the theoretical optimal choice for the level design results in an exponential increase/decrease of performance and the exponential coefficient is a general description metric. For each given autonomy task, we can locate its difficulty and associated theoretical autonomy performance of a free-choice of equipment system, and then extrapolate with the distance of the real-world constraint parameters from the theoretical optimal setup how much better/degraded the system will perform. Since the constraints to be applied instead of a constraint-free-design always have similarities, at least in a specific domain where we always use the same instruments as sensors and actuators, we can store the calculations and the optimal choices in the Enterprise Architecture, enriching it by these entries. And if there are similar patterns between domains, we can store also reference architectures to be quickly re-used in the design phase (or onboard the Field Robot for its local design process).

7.2 Free Will

Free will refers to an individual's ability to make choices and take actions based on their desires, intentions, or beliefs rather than being determined by external factors or pre-existing conditions. It is a concept that has been debated throughout history in various philosophical, religious, and scientific contexts. Free will can be understood as the capacity for individuals to exercise control over their thoughts, decisions, and actions, which implies responsibility and accountability for one's choices. This idea is often contrasted with determinism, where all events are predetermined by prior causes or conditions, leaving no room for individual choice or free will. In the context of philosophy, debates about free will involve questions related to human nature, moral responsibility, and the existence of a higher power or divine intervention that may influence our actions. In science, particularly in neuroscience and psychology, research is focused on understanding how brain processes contribute to decision-making and choice behaviors. In summary, free will represents an individual's capacity for self-determination and autonomy in making choices and taking actions based on their desires or beliefs rather than being dictated by external factors or predefined conditions.

The ethical axiom for the remainder of the discussion is that human beings have a free will. Obviously, the execution of this free will is not unconstraint: Physics generates constraints for the human body and human societies generate constraints for each individuum. Via experience-based learning the generation process of the free will can be affected, but the existence of free will cannot be questioned. This unlimited autonomy feature of human individuals separates them from the Field Robots, we have just designed: The Field Robots stop investigating new levels of detailness according to their programming. By this, the Field Robots generate a predictable environment, which is important for the human to exploit the Free Will by making wishes becoming a reality faster and more efficiently. But the Field Robots designed here deliver also the risk associated to their usage. It is the free choice to ignore the risk, but this is were the human consciousness needs to become active and find the right balance between risk and costs.

7.3 Trust

Trust is a psychological concept referring to the belief in or reliance on someone's integrity, reliability, or ability to perform a particular task or behavior. It involves a sense of confidence that an individual or entity will act in one's best interest and uphold their commitments or promises. Trust plays a crucial role in various aspects of human life, including personal relationships, business interactions, social networks, and political systems. In the context of trust as a mathematical concept, it has been studied using game theory models like the Prisoner's Dilemma and related iterative strategies that involve cooperation between individuals or groups. Trust can be quantified through various metrics such as trustworthiness, credibility, or reliability, which are often used to evaluate an individual's or entity's performance in specific contexts. In summary, trust is a belief in the integrity and dependability of someone or something that affects human interactions and decision-making processes across diverse fields like economics, psychology, sociology, and computer science.

Measuring trust in autonomous systems is an active area of research with no single agreed-upon method. The challenge lies in capturing human perceptions, which are often subjective, uncertain, and context-dependent. One approach to measuring trust involves using surveys or interviews to directly ask people how much they trust a particular system. This can provide valuable insights but may be limited by response bias or lack of consistency across individuals. Another method is to observe behavioral indicators such as willingness to delegate tasks, level of interaction with the system, and compliance with its recommendations. These measures can offer more objective metrics but require careful interpretation and context-awareness. In addition to these direct methods, researchers have also explored indirect approaches that rely on proxies or correlates of trust. For example, one study examined how people's physiological responses (e.g., heart rate, skin conductance) changed when interacting with an autonomous vehicle. Another study analyzed the language used in online reviews to infer levels of trust in a particular brand of self-driving cars. Ultimately, measuring trust in autonomous systems requires a combination of methods that capture both subjective perceptions and objective indicators. The choice of approach will depend on the specific context and goals of the measurement effort. It's important to continuously evaluate and refine these measures as our understanding of human-machine interaction evolves over time.

In the approach for designing Field Robots in this Book, the question of trust goes back to Certification, which is doable: As long as the Field Robot is not progressing with its autonomy beyond the preplanned detailness, no unpredicted emergence can occur, given that the models used for the DSE-M are correct. Trust is therefore refereed back to scientists and engineers, asking the question about how much the models used can be trusted. Therefore, it is important to emphasize that the complete uncertainty represented by uniform distributions used in the DSE-M in this Book seems to be a conservative approach. Also, the maximization of exactitude by automatically increasing the level of detailness is helpful to build trust in the Field Robots. But certainly, with this complete generic approach, trust in autonomous systems is not possible, because calculations cannot be made at an infinite level of detailness which seems to be necessary for an generic approach for any task. Hence, trust needs to know a-priori the limit of necessary detailness, and probably the knowledge of the associated overall task of the environment a Field Robot has to be inserted into.

Take-Away for Design Space Exploration
The technical autonomy recursion stops earlier than the human capacity. The technical systems can have a larger/stronger/more energy hungry/more memory carrying extend, but they miss the right level of profundness as much as the humans do when they are preparing them. The right level of profundness would mean a complete autonomy (a free will), which we as humans cannot provide since we are not able to proof the trustworthyness of such system: Hence, it's just too dangerous to realize a system that has all power, but no trustworthyness. Let's go back to systems where we limit the autonomy to levels at which we as humans can act as the final decider of certification of trustworthyness, and learn by this performing this job how our free will can be made certified to be trustfull. This is not an engineering task, hence naturally we have reached the end of this Book.

Reference

1. McComb, W.D.: Renormalization Methods. Oxford University Press, Oxford (2004)

Chapter 8
Conclusion

An overall procedure capable of utilizing the Unified Architecture Framework (UAF) for automated software generation of Artificial Intelligence Functionalities in robotic applications is presented, consisting of three parts:

1. generating options (given some constraints via the specification/requirement formulations) for entities, services and their combinations,
2. evaluating the specific interface environment for each possible service to mathematically formulate the automated generation of a cost function/potential for an optimization process, and
3. the optimization of this cost function/potential which aims at reaching a maximal independency of demands and abilities with the best fit to the requirements stemming from the operational objective of the system's application, whereby a robust overall system design is maintained through formulating the equilibrium of the relevance of detailedness in the model description as a constraint to the optimization process.

This procedure is generic, takes a SysML-filled UAF as input and generates Artificial Intelligence Functionalities accompanied by a computationally viable validation and verification process, and separates the capacity of the robot's autonomy from the difficulty of given tasks, offering the possibilty for a ranking metric of approaches for implementing autonomous functionalities into robots.

There are several metrics commonly used to measure various aspects of autonomous systems. The metric developed in this Book, although computationally expensive, might help to form a better basis of objectively generated values when comparing autonomous approaches. We add a potential outcome of further elaborations of the results from this Book to the list of commonly used metrics.

F. Ehlers, *Design Space Exploration in Robotics*, Studies in Systems, Decision and Control 229, https://doi.org/10.1007/978-3-031-78740-9_8

- Safety: So far, metrics for measuring the safety of an autonomous system might include the number of accidents or incidents per mile driven, the frequency of unexpected behavior by the vehicle, or the percentage of situations where a human driver had to take over control from the autonomous system. Now, within the Design Space Exploration, we can predict the values, trying to minimize or further mitigate missed detections and/or false alarms resulting in accidents, incidents, or unexpected behavior.
- Performance: So far, metrics for evaluating performance could include measures such as average speed, fuel efficiency, accuracy in following traffic rules, or ability to navigate complex environments. Now, we can optimize in a systematic manner the probably Pareto-surfaces according to operational needs.
- Usability: So far, measures of usability might focus on how easy it is for a user to interact with an autonomous system, including factors like the intuitiveness of its interface, responsiveness to commands, and ability to communicate status updates clearly. Now, we have a handle on how much information we put into the robot in order to minimize redundant communication. Also, we can calculate the relevance of information according to the given task, hence filter out the transmission of irrelevant information.
- Reliability: So far, metrics for assessing reliability could include measures such as mean time between failures, probability of failure in specific scenarios or conditions, or percentage of scheduled maintenance tasks completed on time. Now, we can include the optimization of theses metrics as constraints in the Design Space Exploration.
- Ethics: So far, measures related to ethics might consider factors like how well the autonomous system adheres to ethical principles and values, its ability to make fair decisions that take into account diverse perspectives, or the extent to which it is transparent about its decision-making processes. Now, we can strictly add ethical constraints to the Design Space Exploration.
- Acceptance: So far, metrics for gauging user acceptance could include measures such as willingness to use an autonomous system in various situations, level of trust in the system's capabilities, or overall satisfaction with the experience of using the system. Now, we can correlate objectively generated values from the Design Space Exploration with the subjectively recorded metrics. Furthermore, with the prediction capability of Design Space Exploration, the responsibility can be handed from the Designer to the User with naming and objectively rating the associated risks.
- Sustainability: So far, measures related to sustainability might focus on factors like energy efficiency, environmental impact, or ability to operate without significant maintenance or repair for extended periods of time. Now, architectures and Digital Twin methodologies can be used to optimize also for sustainability. Furthermore, and most important from a theoretical physicist's point-of-view, based on the

understanding of the principles behind the Design Space Exploration there is a fixed starting point of adjustments and optimization: Invention time, the numbers of repeated simulations and data intensive learning loops in artificial intelligence can all be lowered, since previous results are stored in the architectural environment associated with the automated code generation procedures in the Design Space Exploration-Machine.

Glossary

This glossary tries to focus on the usage of the words related to the topics and perspectives presented in the book. This glossary does not aim for giving definitions, it rather explains the context in which they are used in the book.

Abstraction Abstraction refers to the process of simplifying complex concepts or systems by representing them using more generalized models, ideas, or notations that capture their essential characteristics while hiding unnecessary details: In computer science—Abstraction enables developers to create reusable and modular code components by isolating specific functionalities from a larger system, making it easier to understand, maintain, and modify. In mathematics—Abstract algebra is a field that studies general concepts like groups, rings, fields, and vector spaces, allowing mathematicians to unify diverse areas of study under common principles. In philosophy—Abstraction helps philosophers to analyze complex ideas or phenomena by breaking them down into simpler components or propositions, facilitating logical reasoning and critical thinking. In summary, abstraction is a powerful concept that simplifies the understanding and manipulation of complex systems by representing them using generalized models or notations that focus on their essential characteristics while hiding unnecessary details.

Autonomy Autonomy refers to the capacity for a system or entity to function and make decisions on its own, without direct external control or intervention. In the context of technology, autonomous systems are designed to perform tasks independently by processing information, making decisions, and taking actions based on predefined rules, algorithms, or machine learning techniques: Autonomous vehicles—Cars, drones, or robots that can navigate their environment without human input. Robotic process automation (RPA)—Software bots designed to perform repetitive tasks within an organization's existing IT infrastructure. Artificial intelligence (AI) systems—Machine learning algorithms and neural networks capable of adapting, learning, and improving their performance over time. In summary, autonomy is the

F. Ehlers, *Design Space Exploration in Robotics*, Studies in Systems, Decision and Control 229, https://doi.org/10.1007/978-3-031-78740-9

ability for a system or entity to operate independently by making decisions and taking actions based on available information without requiring constant human input or control.

The levels of autonomy refer to a hierarchical framework that classifies the degree of self-direction or decision-making capabilities in autonomous systems: Level 0 (Manual Control): The system entirely relies on human intervention for all tasks and decisions, with no automation involved. Level 1 (Driver Assistance): The system provides assistance to the user by offering suggestions or aiding specific functions but still requires continuous human supervision and control over critical decisions. Level 2 (Partial Automation): The system can perform some tasks autonomously, such as maintaining speed or lane position in vehicles, while requiring occasional human intervention for complex situations. Level 3 (Conditional Automation): The system is capable of performing all necessary functions without direct user input but still requires the driver to be prepared to take control when needed, such as during emergencies or extreme conditions. Level 4 (High Automation): The system can operate autonomously in specific environments and situations, with no requirement for human intervention unless desired by the user. Level 5 (Full Autonomy): The system is capable of operating without any need for human input or supervision, functioning independently in all conditions and scenarios. In summary, levels of autonomy are a way to classify autonomous systems based on their ability to perform tasks independently or with varying degrees of human involvement, ranging from minimal assistance (Level 0) to complete independence (Level 5).

Axiom An axiom refers to a self-evident truth or a foundational principle accepted as a matter of course without requiring proof or demonstration. Axioms serve as the basis for logical reasoning, deductive arguments, and mathematical theories: In mathematics—Axioms are fundamental postulates that form the foundation of a particular branch of math, allowing more complex theorems to be derived through formal logic and proofs. In philosophy—Philosophical axioms or first principles provide the starting points for building philosophical systems and theories by establishing basic assumptions about reality, knowledge, ethics, or existence. In computer science—Axiomatic design is an approach to system development that uses a set of axioms as guiding principles to ensure consistency between requirements, specifications, and implementations. In summary, an axiom is a fundamental principle or self-evident truth that serves as the basis for logical reasoning, deductive arguments, and theoretical foundations in various domains such as mathematics, philosophy, and computer science.

Blackbox Using a piece of hardware or software without knowledge about its internal structure. A black box is an enigmatic device or system whose inner workings are concealed from external observers, focusing on input/output relationships rather than the mechanisms governing its behavior.

Concept A concept is an abstract idea or mental representation that helps humans understand, categorize, and communicate information about the world around them. Concepts are formed by grouping together related ideas, experiences, or observations to create a simplified understanding of complex phenomena. In various fields such as philosophy, psychology, linguistics, and artificial intelligence, concepts play an essential role in organizing knowledge, facilitating learning, and enabling effective communication. A "concept" in C++ refers to a feature introduced in the language that helps programmers define requirements for template arguments (i.e., types or values used with templates). Concepts provide a way to ensure that template functions and classes are only instantiated with suitable input parameters, improving code correctness and maintainability. A concept declaration defines a set of constraints on the template arguments, specifying their required properties or behaviors. When using a concept in a function or class definition, the compiler checks if the provided template argument satisfies these requirements at compile time. If it does not meet the criteria, the programmer receives an error message and is prompted to fix the issue before proceeding with compilation. Concepts are still under development for C++20 and may change in future language revisions. However, they aim to improve generic programming by providing a more robust way of ensuring that templates work correctly with their intended input parameters.

Control In robotics, the distinction between planning and control lies in their respective roles within the overall robotic system architecture: Planning: Robot planning involves determining high-level goals or strategies for a robot to achieve its tasks. This process typically consists of several stages: problem definition, state space representation, cost function formulation, search algorithms, and solution evaluation. The output from this stage is often a sequence of actions (or trajectories) that the robot should follow to accomplish its mission. Control: Robot control refers to the real-time execution of low-level commands based on the high-level plans generated by planning. This process involves converting the planned trajectory into specific motor commands, adjusting for feedback from sensors and actuators, and ensuring that the robot's physical constraints are respected during its motion. Control systems in robots often employ techniques such as PID controllers (Proportional-Integral-Derivative), optimal control methods, or machine learning algorithms to regulate the robot's behavior in real time. In summary, planning focuses on generating a suitable plan for achieving a goal, while control is responsible for executing this plan by adjusting the robot's actions based on feedback and constraints during operation.

Digital Twin A digital twin is a virtual counterpart that mirrors its physical entity or system by continuously learning from data inputs, simulations, and analysis to provide valuable insights for optimization, monitoring, and decision-making purposes.

Enterprise Architecture An enterprise architecture (EA) refers to a comprehensive framework that outlines the structure, principles, and processes governing an organization's IT systems, applications, data management, and business operations: Strategic alignment—EA ensures that technology investments are aligned with organizational goals and objectives for improved efficiency, productivity, and competitiveness. Standardization—It promotes the use of consistent tools, methodologies, and practices across different departments or divisions to simplify collaboration, reduce redundancy, and minimize risks. Governance and compliance—EA provides a structured approach for managing IT resources, ensuring adherence to regulatory requirements, industry standards, and best practices. Agility and scalability—It enables organizations to adapt quickly to changing business needs or market conditions by facilitating the integration of new technologies, systems, or processes into existing infrastructure. In summary, an enterprise architecture is a strategic approach that helps organizations plan, design, and manage their IT environment in line with their overall objectives, ensuring optimal performance, efficiency, and flexibility across all aspects of the business.

Field Robotics Field robotics is an interdisciplinary domain that focuses on developing robotic systems capable of operating in unstructured outdoor environments, addressing challenges such as navigation, communication, autonomy, and interaction with the natural world for various practical applications.

Generative Programming Generative programming is an approach to software development that focuses on creating reusable components and modules rather than writing code for specific tasks or problems. It involves designing and implementing generalized functions, classes, or interfaces that can be easily adapted or configured to solve a variety of related issues. This method aims to reduce the time and effort required in developing new applications by leveraging existing code elements and promoting modularity, flexibility, and maintainability within software systems.

Generic Programming Generic programming refers to a programming style or paradigm where functions, classes, or interfaces are designed in such a way that they can be applied to different data types without the need for modifications. This approach promotes code reuse and maintainability by allowing developers to create generalized solutions that can adapt to various specific situations. Some key concepts of generic programming include: Templates: In languages like C++, templates allow programmers to define functions or classes with placeholders for types or values. When a template is instantiated (used in the code), these placeholders are replaced by actual data types or values, creating specialized versions of the generalized component. Generics: Languages such as Java and C# provide built-in support for generics, which allow developers to create parametrized classes or methods that can work with different data types without needing casts or type checks. This improves code safety and readability by enforcing strong typing rules. Polymorphism: In object-oriented programming (OOP), polymorphism allows a single function or method to take on multiple forms based on the context in which it is used. This can be achieved through inheritance, where subclasses override parent class methods with specialized implementations, or through parametric polymorphism using templates/generics. Functional Programming: In functional programming languages like Haskell and Lisp, higher-order functions and pure functions are used to create generic algorithms that can be applied to different data types without modification. This approach promotes code reuse by abstracting common patterns into generalized components. Algebraic Data Types (ADTs): In languages like Haskell or OCaml, ADTs provide a way to represent complex data structures as sum and product types. These types can be used in generic functions to ensure strong typing rules and improve code safety.

Global Optimization of Design In Global Optimization of Design the intend is to optimize for the strongest requirement and widest specification possible. It uses the top-down design approach with a given set of requirements and specifications, then asking the question which most difficult requirements can be fulfilled with as wide as possible specification. In a market scenario, product design is one example where the best product is aimed for. Another example is evolution where the most versatile autonomous being can be seen as a product of a continuous selection process.

Legendre Polynomials Legendre polynomials are orthogonal because they satisfy the Legendre differential equation, which relates their coefficients to their derivatives. Specifically, the Legendre differential equation states that: $d/dx(Pn'(x)) = n * Pn(x)$ where Pn is a Legendre polynomial of degree n and Pn' is its derivative. The orthogonality relation implies that any linear combination of Legendre polynomials can be expressed as a weighted sum of the individual Legendre polynomials. This property makes Legendre polynomials useful in many applications, such as solving differential equations and approximating functions.

Machine A machine is just doing what it has been programmed for.

Ontology An ontology refers to a formal representation of knowledge and concepts within a specific domain or area of interest. It aims to define the terms, relationships, and rules governing the entities in that particular field, enabling better understanding, communication, and reasoning about the subject matter. Ontologies can be applied across various domains such as artificial intelligence, natural language processing, information retrieval, knowledge representation, and more. Key components of an ontology include: Concepts (Classes or Categories)—These are the primary building blocks that represent entities within a domain, such as objects, events, properties, or relationships. Relationships (Properties or Attributes)—These define how concepts relate to one another and help establish connections between different elements in an ontology. Axioms or Rules—These are statements that express general principles or constraints governing the entities within a domain, providing a basis for logical reasoning and inference. Instances (Individuals)—These are specific examples of concepts that illustrate how they apply to real-world situations or instances. In summary, an ontology is a structured framework that represents knowledge about a particular subject area by defining its key entities, relationships, rules, and constraints, facilitating communication, understanding, and reasoning within the domain.

Probability Probability refers to the likelihood or chance that an event will occur in a given situation. It is used to quantify uncertainty, expressing how likely it is for something to happen compared to other possible outcomes. In mathematics and statistics, probability is represented as a number between 0 and 1, where 0 indicates no chance of the event occurring, and 1 signifies certainty that the event will take place. Probability theory provides a formal framework for understanding and calculating probabilities in various situations, such as flipping coins, rolling dice, or drawing cards from a deck. It employs concepts like conditional probability, independence, and Bayes' theorem to analyze complex events and update beliefs based on new information. Probability is widely used in fields like statistics, machine learning, finance, gaming, and scientific research for making informed decisions under uncertainty.

Process A process is a structured framework that outlines the actions, steps, or procedures necessary to achieve specific objectives or goals within a given context or situation.

Reference Architecture A reference architecture refers to a standardized, well-defined structure or blueprint for designing and implementing software systems, hardware solutions, or other technology-based infrastructures. This architectural framework serves as a foundation for developers, engineers, and IT professionals to build upon when creating new applications, products, or services within specific domains or industries. Reference architectures often consist of multiple layers, components, and interfaces that facilitate the integration of various technologies, tools, and methodologies into cohesive systems. They provide a common language for stakeholders to discuss system requirements, design decisions, and potential improvements while ensuring compatibility with existing standards and best practices. By adopting reference architectures, organizations can streamline their development processes, reduce costs, and improve overall efficiency in delivering technology-driven solutions.

Robot A robot is an automated machine or system designed to perform tasks and functions typically associated with living beings, such as sensing, perceiving, adapting, learning, or manipulating objects in its environment. Robots can be found across various industries, including manufacturing, healthcare, agriculture, space exploration, military operations, and entertainment. Robots are often classified into different categories based on their design, functionality, and purpose: Industrial robots: These are designed for repetitive tasks within a controlled environment, such as welding, painting, or assembly lines in manufacturing facilities. They typically follow pre-programmed paths and have limited sensing capabilities. Service robots: Designed to assist humans with daily tasks, service robots can be found in domains like healthcare (e.g., medical assistance), hospitality (e.g., concierge services), or domestic settings (e.g., vacuum cleaners). They often incorporate advanced sensing and navigation systems for interacting safely within human environments. Military/defense robots: These are engineered to perform tasks in hazardous conditions, such as reconnaissance, bomb disposal, or surveillance. They may be remotely controlled or autonomously operate using AI algorithms and sophisticated sensors. Entertainment/educational robots: Designed for amusement or educational purposes, these robots can interact with humans through games, performances, or simulations. Examples include humanoid robots in theme parks or social robotics projects aimed at fostering communication between humans and machines. Scientific/exploration robots: Used to study extreme environments or locations that are otherwise difficult for humans to access, these robots can be found in fields like space exploration (e.g., Mars rovers), underwater research (e.g., oceanographic probes), or environmental monitoring (e.g., weather stations). Biological/biomimetic robots: Inspired by living organisms, these robots often incorporate organic materials and mechanisms to replicate the functions of natural systems, such as self-healing, swarm intelligence, or adaptive behavior.

Spiral Development Spiral development is seen as a Bottom-Up design approach with periodic adjustment regarding development goals. In this sense, spiral development does not point to a strict given design goal, but explores how far one can get starting with a given set of setup conditions. Spiral development is to be contrasted with Global Optimization of Design where the intend is to optimize for the strongest requirement and widest specification.

Strategy A strategy refers to a plan or approach that outlines the objectives, goals, and actions required for achieving desired outcomes within a specific context or situation. Strategies can be applied across various domains such as business, military, politics, personal development, and more. They typically involve: Identifying key challenges or opportunities—Understanding the current state of affairs and recognizing areas that require attention or improvement is crucial for formulating an effective strategy. Setting clear objectives and goals—Establishing specific targets and milestones helps provide direction, focus, and motivation to those involved in executing the plan. Analyzing available resources—Assessing the human, financial, technological, or other resources at one's disposal is essential for allocating them efficiently towards achieving strategic objectives. Developing actionable steps—Creating a detailed roadmap of tasks and activities that must be carried out to reach the desired outcomes ensures a structured approach to strategy implementation. Monitoring progress and adjusting as needed—Regularly reviewing progress, evaluating results, and making necessary modifications or adaptations is vital for staying on track towards achieving strategic goals. In summary, a strategy serves as a comprehensive roadmap that outlines the objectives, actions, and resources required to navigate complex situations or challenges effectively while ensuring desired outcomes are met successfully.

Technical Robot A Technical Robot is white-boxing a given ontology with a given reference, i.e. it is not capable to develop own ontological elements by inspecting the environmental processes around it. Contrast this with a Field Robot which can handle emergent situations.

Validation Validation refers to the process of ensuring that a product or service meets specific requirements, standards, or criteria set by stakeholders, regulatory bodies, or users. This process typically involves testing and verifying the quality, safety, efficiency, and effectiveness of an offering in real-world scenarios to identify any potential issues, risks, or areas for improvement. In software development, validation ensures that a product meets user needs, technical specifications, and legal requirements before being released into the market. This process may involve various stages such as requirement gathering, design review, prototyping, testing, and feedback collection to identify any discrepancies between the intended functionality and actual performance. In other industries like healthcare or manufacturing, validation can encompass a wide range of activities, including quality assurance, risk management, process optimization, and continuous improvement initiatives. The ultimate goal is to ensure that products and services are safe, reliable, and fit for their intended purpose while minimizing risks and costs associated with non-compliance or failure.

Verification Verification refers to the process of confirming that a product or service adheres to its design specifications, technical requirements, and functional criteria throughout its development lifecycle. This process aims to identify any discrepancies between intended functionality and actual performance, ensuring that the final output meets quality standards and stakeholder expectations. In software engineering, verification involves various activities such as code reviews, unit testing, integration testing, system testing, and acceptance testing to ensure that each component of a project functions correctly and interacts seamlessly with other parts of the system. This process helps identify bugs, security vulnerabilities, performance issues, or compatibility problems early in development, allowing for timely corrections and improvements before deployment. In other industries like engineering, construction, or finance, verification may encompass inspections, audits, assessments, or certifications to ensure that products, services, or processes comply with industry standards, regulatory guidelines, and best practices. The ultimate goal of verification is to minimize risks, maintain quality, and foster confidence in the final output among stakeholders, users, and regulators.

Whiteboxing Whiteboxing refers to the process of implementing a software component or module without any consideration for its external interfaces, dependencies, or interactions with other parts of the system. This approach focuses on the internal design and functionality of an individual unit rather than its integration within a larger architecture. In whitebox testing, developers focus on testing each component individually to ensure that it meets specified requirements and works as intended when isolated from the rest of the system. Whiteboxing is particularly useful during software development for identifying bugs, optimizing performance, or refactoring code without being constrained by external factors such as API compatibility or integration with other modules. In contrast, blackbox testing emphasizes evaluating a component's behavior and functionality from an end-user perspective, considering its interactions with the rest of the system but not worrying about the internal workings.

The manufacturer's authorised representative in the EU is Springer Nature Customer Service Centre GmbH, Europaplatz 3, 69115 Heidelberg, Germany. If you have any concerns regarding our products, please contact ProductSafety@springernature.com

Printed and bound by CPI Group (UK) Ltd, Croydon, CR0 4YY
30/09/2026
02225574-0003